QH 315 .B32 1970
Baker, Bill, 1942-
Modern lesson plans for the
biology teacher

T5-CAE-255

ARNULFO L. OLIVEIRA MEMORIAL LIBRARY
1825 MAY STREET
BROWNSVILLE, TEXAS 78520

Modern Lesson Plans for the Biology Teacher

BILL BAKER

St. John's University

and

HELEN HOCH KOTSONIS

Jersey City State College

PARKER PUBLISHING COMPANY, INC.

West Nyack, N.Y.

ARNULFO L. OLIVEIRA MEMORIAL LIBRARY
1825 MAY STREET
BROWNSVILLE, TEXAS 78520

© 1970 BY

PARKER PUBLISHING COMPANY, INC.

WEST NYACK, N.Y.

ALL RIGHTS RESERVED. NO PART OF THIS
BOOK MAY BE REPRODUCED IN ANY FORM OR
BY ANY MEANS, WITHOUT PERMISSION IN
WRITING FROM THE PUBLISHER

LIBRARY OF CONGRESS
CATALOG CARD NUMBER: 70–86535

Second Printing.....July, 1971

PRINTED IN THE UNITED STATES OF AMERICA

13–594986–6 B&P

To John Gabriel Navarra

A Sourcebook for the Biology Teacher

This book was written to fill the need for a concise and well organized collection of lesson plans and materials for the biology teacher, particularly at the secondary school level.

At present, there are many sourcebooks available to the biology teacher. It has been our experience, however, that each of these carries only bits and pieces of information, which must be sought out and organized. They require the availability of numerous different sources together with extensive knowledge on the part of the teacher as to where to find these materials. To the beginning teacher this is a limiting factor and a handicap; to the experienced teacher this is a misuse of valuable time—time that could be used more constructively in the classroom.

This book can be *used by all who teach biological science.* For beginning teachers and those with only a few years experience, this book will prove a handy, reliable, and practical source for the planning, preparation, and execution of effective lessons. For the experienced teacher it will serve to provide new ideas and new methods for presenting and teaching the subject, while at the same time allowing him to use the lessons as building blocks or as a framework for further expansion or amplification, as time permits and as experience justifies.

Our aim is to provide a book which is practical in nature, offering the secondary school teacher concrete activities, lessons, materials, and information specifically geared to making his work more effective and efficient. In other words, we are striving to supply the teacher with a guide that is applicable to the classroom situation, a book which will provide ideas about dealing with the basic materials of secondary school biology, and ways to stimulate student understanding and participation.

We have included those areas of biology that are usually not stressed in the sources readily available to the teacher, and which are a major part of the curriculum.

LESSON PLAN STRUCTURE AND APPLICATION

Integrated lessons. One of the unique aspects of this book is that it is thoroughly integrated. *Complete lessons*, including all auxiliary materials—suggested time needed for preparation and lesson coverage, needed materials and equipment, specimens to be used and their source—are to be found in one place. There will be little time wasted by the teacher in searching through several scattered sources. References are made to those areas which may prove difficult, together with suggestions for alternate presentations and demonstrations which will help to clarify the concepts presented.

Structured lessons. The book has been structured so that the teacher may select single lessons or a *group* of related lessons which may be used in any order he chooses in keeping with the organization of the biology curriculum. It is suggested that the teacher select the appropriate lessons in advance and read through them carefully. It will be noted that an approximation of the time needed for presentation is indicated at the beginning of each lesson, to the right of and slightly below the title. Detailed instructions for laboratory procedures and time needed for preparation have been indicated where appropriate. Certain procedures have been included which will serve as valuable laboratory experiences for students, while others have been identified as demonstrations best used during lecture-discussions.

Concepts developed. Each lesson develops several related concepts. Alternate methods of presentation have been included so that a number of different approaches will be available to the teacher. This choice of methods allows for development of concepts at different levels of understanding. It may be advantageous to use several alternate methods for the development of more complex ideas. In the lesson are student directed questions and, as a summation of the concepts and ideas developed, pertinent facts have been listed where appropriate.

Diagrams, quiz. Illustrations and diagrams have been included which should prove useful not only in developing suggested demonstrations and laboratory procedures, but also as audio-visual aids to be used with the chalkboard, opaque, and overhead projector. The diagrams have been kept simple so that they can be easily reproduced by the teacher on the chalkboard. A quiz has been added to the end of each lesson. The questions have been designed to motivate students to apply what they have learned. This is intended as a framework which can further be developed by the teacher.

Readings, films, models. A short list of readings has been offered as supplementary material. The readings for each lesson have been kept to a minimum to encourage both teacher and student use. Also included is a list of films which can be used to supplement and further develop the lesson. Film source, address of distributor, and rental information are indicated. Student interest can be developed further through the use and/or construction of models and charts. Various types of student constructed or commercially prepared materials have been proposed.

Materials tested. It should be pointed out that all lessons, demonstrations, and experiments were tested and used successfully in our classrooms. Students reacted with enthusiasm to the demonstrations, laboratory experiences, and films, and particularly enjoyed the many student-oriented projects. As a result of this successful use of materials, we are able to offer specific helpful suggestions throughout, which should prove to be useful to the teacher in presenting the lesson.

Included in each completed lesson will be:

1. appropriate demonstrations and experiments dealing with the unit to be studied;
2. laboratory procedures for students;
3. suggested laboratory preparations, including equipment and time requirements for the teacher;
4. alternate methods of presentation to help clarify the lesson's difficult concepts;
5. appropriate illustrations and diagrams to help clarify the lesson;
6. pertinent facts to be emphasized;
7. suggested questions for a possible quiz;
8. list of audio-visual materials and models appropriate for that particular lesson;
9. suggested bibliography for *both* teacher and students.

We are grateful to Cyrus Barnes, Stephen Ferko, George Mitchell, and Herbert Schwartz who have served as sources of continuing inspiration for us. To these master teachers, we are forever indebted.

Our special thanks to Dean Baker for his help in the preparation of the manuscript, and to George Kotsonis for his patience and understanding.

The Authors

CONTENTS

Unit I: The Laboratory and Its Uses • 17

A unit designed to familiarize the student with laboratory techniques and procedures, including the proper use and care of the microscope, as well as other standard laboratory materials. Students are guided in developing creative and meaningful laboratory procedures with emphasis on data collection, careful planning, and investigation.

Lesson 1: The Proper Use of Materials • 18

An introductory laboratory lesson which stresses the proper handling of laboratory materials. Included are suggestions concerning the use of liquid and dry chemicals, as well as living and preserved materials.

Lesson 2: Care and Use of the Microscope • 25

Designed to introduce fundamental techniques for the use and care of the microscope as well as to develop a fundamental understanding of the way the microscope works.

Lesson 3: Preparation of Permanent and Temporary Slides • 32

Basic methods for making simple permanent slides, as well as temporary wet-mount slides, are explored.

Lesson 4: Looking at Cells and Judging Their Size • 38

An introduction to and look at microscopic measurement and use in observing and describing cells.

Lesson 5: The Development of Controlled Experiments • 41

An explanation of the scientific method as it relates to the actual design and implementation of an experiment.

Unit II: The Cell • 45

More and more biology courses center around the cell and its functions. This unit includes a series of lessons dealing with the structure and function of the generalized cell.

Lesson 6: The Cell Membrane and Molecular Movement • 46

A look at how substances move into and out of the cell in relation to the basic principles of diffusion-osmosis. A brief exploration of the double lipo-protein nature of the accordion-pleated cell membrane. Basic demonstrations include the use of various membranes to explore the nature of different types of solutions. Special emphasis is placed on the hypertonic-hypotonic-isotonic solutions in relation to a perfect osmotic unit and their effects on living cells.

Lesson 7: Generalized Cell: Internal Structure and Function • 55

A look at the structure and function of a typical cell, excluding the nucleus and including an exploration of the differences between plant and animal cells.

Lesson 8: The Genetic Code • 61

A biochemical study of the nucleus, including the more recent information on DNA and RNA, ending with the genetic code. Basic illustrations include use of the Watson-Crick model of DNA. Some suggestions for the use of simplified charts explaining the construction of the genetic code are included.

Lesson 9: Cell Division • 66

A clarified study of mitosis and meiosis serving as a consolidation of concepts developed in Lessons 7 and 8. Includes the use of root tip slides and instructions for the use of onion root tips by students in the preparation of slides for the study of mitotic principles.

Lesson 10: Heredity • 76

The genetic code is used as the basis for the development of the basic concepts of the transmission of hereditary characteristics from generation to generation, working from the cell to man. Included is the use of 35mm color slides to graphically show Mendelian genetics. Instructions are included for the production of slides by the teacher, if not readily available.

Unit III: Plants: Form and Function • 87

As the practice of most biology curricula is to stress the basic parts of the plant in the handling of the subject matter, this unit, therefore, presents a concise cross-sectional representation of lesson plans dealing with the various functional aspects of plant structure.

Lesson 11: The Root and Its Functions • 88

An illustration of the conductive property of roots, with structural implications which can be used in conjunction with Lesson 12. Describes how water and other materials are translocated to the various plant parts with the aid of root pressure, cell sap, and gravitational effects.

Lesson 12: The Stem and the Conduction of Fluids • 97

A capsule lesson dealing with a further illustration of conductive properties introduced in Lesson 11, with structural implications of the conductive functions of the stem. Use of living specimens can be made to show stem anatomy, highlighting the major conducting organs, the fibrovascular bundles.

Lesson 13: The Leaf and Food Making • 104

A study of the structure of the leaf in relation to photosynthesis and food production of the plant as a whole. Experiments include chlorophyll extraction and chromatography.

Lesson 14: The Leaf and Respiration • 112

A study of gas exchange as a further development of the varied aspects of photosynthesis, with a look at stomata and their functions.

Lesson 15: Flowers, Fruits and Seeds • 116

The development of a comprehensive lesson which allows for progression from the simple to the complex concepts of the various aspects of the reproductive cycle of flowering plants. Extensive use is made of living specimens. A series of student-prepared germination demonstrations are included.

Lesson 16: Plant Responses • 123

An integrated study of many phases of plant reactions encompassing the mechanisms involved in growth and turgor movements, together with the influence of plant hormones. Demonstrations include using gibberellins, auxins and their effects on plant responses, as well as activities on phototropism, photoperiodism, geotropism, hydrotropism, and thigmotropism.

Lesson 17: Ecology and the Terrarium • 134

The study of various ecological principles through the use of a teacher-planned and student-developed terrarium. The terrarium may be used by the teacher as a focal point in stimulating student interest and in developing related concepts. This provides an excellent opportunity for observing the interrelationships between plants and animals in their simulated environments.

Lesson 18: Insect-Eating Plants • 142

A capsule lesson presenting some interesting materials concerning the observation, cultivation, and behavior of insectivorous plants.

Unit IV: The Organ Systems of Man • 149

A series of lesson plans basic to the fundamental concepts of form and function as explored through an in-depth study of the organ systems of man.

Lesson 19: Locomotion and Support • 150

An integrated lesson centering on the skeletal and muscular systems. Included is the use of the human skeleton, fresh long bones, prepared slides of bone and muscle. Dissections and laboratory preparation of small animal skeletons are included.

Lesson 20: The Digestive System and Its Functions • 164

The physical and chemical breakdown of foods, their absorption and storage, and the elimination of the solid wastes. Laboratory activities include taste tests and chemical studies of the breakdown of food groups such as protein tests, as well as carbohydrate and fat studies.

Lesson 21: The Excretory System and Its Functions • 175

A study of the parts of the excretory system of man and of their functions in the formation and elimination of liquid wastes. Laboratory activities include urea tests and optional material on urine analysis.

Lesson 22: Respiration in Man • 181

A study of the exchange of gases between the lungs and the blood and cells. Included is the use of a balloon and Y-tube in a bell jar to demonstrate pressure changes in the chest cavity during breathing. Also illustrated are the various student-centered tests for lung capacity, breathing rate, and related factors.

Lesson 23: Blood and Circulation • 187

A lesson culminating the study of the heart and circulatory system of representative animal types begun in Lesson 23, with a look at the circulatory system and four-chambered heart of man. Laboratory activities include the dissection of a sheep heart, as well as blood groups and blood typing.

Lesson 24: The Endocrine System • *197*

A look at the various endocrine glands and their functions in man. The lesson includes normal anatomy as well as various abnormal conditions.

Lesson 25: Nervous and Sensory Mechanisms • *202*

Discussion includes the anatomy of the nervous system of man and its functions. Included is a study of the nervous system as the sensory system of the body. Illustrated with a series of student-conducted tests, including the knee-jerk reflex, oculomotor reflex, and cutaneous senses.

Lesson 26: Homeostasis • *217*

A look at the integrated functioning of the human body through the interaction of the excretory, circulatory, respiratory, endocrine, and nervous systems.

Lesson 27: Reproduction and Development • *221*

An investigation into the biological origins of man. The lesson includes a discussion of the anatomy and physiology of the egg and sperm cells, their origins, fertilization, embryological and fetal development. Extensive use is made of audio-visual materials.

Unit V: The Origin and Evolution of Living Things • 229

A unit dealing with evolution which covers theories ranging from the historical to the contemporary. Also included is the development from single cell to multicellular organism.

Lesson 28: The Origin of Life • *230*

An introduction to the various theories dealing with the beginnings of life, commencing with a historical review and concluding with current biochemical principles.

Lesson 29: From the Simple to the Complex—Evolution • *235*

The evolution of man and other organisms, as well as that of plants, is discussed.

Unit VI: Drugs and Addiction • 239

A unit dealing with this vital issue in modern society which should not be ignored in current biology curricula. This unit contrasts the great improvements in our lives that resulted from the development of drugs, with the tragedy of the wanton misuse of drugs.

Lesson 30: The Medical Uses of Drugs • *240*

A specific study of various classes of drugs and their great importance to health when used properly.

Lesson 31: The Effects of Drug Misuse • *245*

A presentation of the facts concerning the harmful effects of improper drug use as related to the human organism. Suggestions for the use of outside sources of information have been included.

Index • 250

Unit I

THE LABORATORY AND ITS USES

Lesson 1

THE PROPER USE OF MATERIALS

Laboratory time: 45 minutes

AIM

To introduce the student to the proper methods in the care and handling of standard laboratory materials.

MATERIALS

Filter paper; Simple balance; Film; Graduated cylinder.

PLANNED LESSON

1. A look at the laboratory

To get your students off on the right foot concerning the handling of various materials in the laboratory, it usually is a good idea to familiarize them with the physical aspects of the laboratory on the first day of class. At that time, the teacher can acquaint his class with the various equipment and materials and familiarize them with their use.

2. Proper handling of chemicals

A. Dry Chemicals

As with liquids, students must be aware of the need for care so as not to contaminate dry chemicals. They should be taught a standard procedure for the use of dry materials, such as removing the material from a small labeled beaker *only* after it has been prefilled by the teacher or laboratory assistant. Only the approximate amount needed should be taken, as no materials should be returned to a supply source once they have been removed. Students should be asked to explain the reason for this.

For the proper transport of materials, students should use a piece of filter paper which they folded in half, opened flat, and then folded in half again at right angles to the original fold. This tends to make a slight depression in the center of the paper and prevents material from falling off. (To prevent an uneven surface, all creases should be made on the same side of the paper.) As students approach the table on which all materials have been neatly arranged, they should select the one needed, carefully check the label, and gently shake out enough material onto the filter paper. It is sometimes helpful to rotate the beaker in order to loosen the contents. Care should be used to avoid spillage. Any material which is spilled should be cleaned up by that student immediately. (It is vital to train every student to clean up after himself—so that the laboratory will remain clean and in proper order, as well as to prevent accidents—and to train him to accept responsibility for his own actions.) Special instructions should be given *before* the laboratory begins for any chemicals which may need special handling.

At this point you should explain the use of a simple balance, and may also wish to explain the use of the analytical balance to more advanced groups. A worthwhile film for this purpose is "Principles of the Analytical Balance."

B. Liquid Chemicals

Special care must be taken with liquid chemicals. Pouring should be done slowly, gently, and carefully. Obviously, the care of acids is a most important point to be covered. Again, it would be easiest if the teacher were to transfer small quantities of the acids to labeled beakers.

Students must then be instructed in the proper use of the graduated cylinder. They should be reminded to read the liquid level in the cylinder at the center of the fluid level (the meniscus). Figure 1-1 can be used to illustrate this point. Students must be cautioned always to pour acid into water when such a mixture is needed, *never* water into acid. If you point out that a great deal of heat is released when acid and water mix, students should be able to understand that a sudden production of gas and splattered acid results from the addition of a small amount of water as opposed to the harmless dissipation of heat if a small amount of acid is added to a larger amount of water.

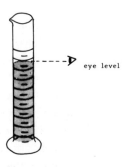

Figure 1-1. Proper Reading of the Graduated Cylinder.

STUDENTS MUST ALWAYS WEAR PROTECTIVE GLASSES WHEN WORKING WITH CHEMICALS. It is wise to require a laboratory coat or apron to protect clothing as well. In order to prevent accidents, careful instructions should always precede an exercise requiring the use of chemicals. In order to prevent injury if an accident should occur, students should be carefully trained in emergency procedures before beginning work. They should be told to bathe any area which may have been touched by any chemical, especially acids, in clear running water. In the case of acids, a weak solution of a base might be prepared by the teacher in advance and kept in a prominent location which has been pointed out to students. (A weak solution of sodium hydroxide can be used.)

C. Do's and Don'ts in the Laboratory

A number of general laboratory techniques should be emphasized to the students in advance. Included should be:

   1. Labels should be read carefully; students should label bottles and beakers so that they know their contents.

2. Supply bottles should never be removed from their central location.
3. A heated test tube should never be pointed at anyone.
4. Care should be exercised in the handling of hot glass.
5. Proper techniques should always be used when mixing acid and water.
6. To prevent accidents the teacher's instructions should always be carefully followed.
7. Students should be aware of the location and use of fire extinguishers.
8. Each student should know where to find fire blankets, safety showers, and first aid kits, and how to use them.
9. It is essential that the location and use of all safety equipment be explained to students during this first orientation laboratory session.
10. Students should not be allowed to devise and perform experiments without teacher approval.

D. Proper Disposal of Materials

Students should be instructed concerning the proper disposal of unused chemicals and waste materials. It is generally advisable to have special garbage cans for glassware and dry chemicals. These should be pointed out to the janitorial staff so that they can be properly and harmlessly destroyed. Liquids may be washed down a laboratory sink if the following procedure is followed: Allow clear water to run down the drain for at least 10 seconds at strong pressure. Turn off the water, pour a small amount of liquid *directly into the drain* (not the sink) and gently wash down with water. Repeat this procedure until the fluid is completely eliminated. In the case of acids, neutralization should be affected first. (Before undertaking any laboratory procedure involving such materials, it is advisable to consult with the chemistry teacher, or if not available, janitors or others who are familiar with your school's construction and procedures. There are often special disposal equipment and/or procedures already established which should be followed.)

3. Living and preserved materials

A. Living Materials

The maintenance of living materials depends upon providing optimum conditions of survival. Generally, the use of such materials in the laboratory requires as little disturbance of them as possible, in order that normal structure and function can be observed. Students must be as gentle as possible with both plants and animals. The teacher should contact his local state department of health for information concerning the proper housing and maintenance of living organisms in the laboratory. Living materials commonly studied in the laboratory would include cultures of protozoa, frogs, worms, insects.

Protozoa

These can be ordered from commercial supply houses such as: General Biological Supply House, Inc., 8200 South Hoyne Avenue, Chicago, Ill. 60620; or Carolina Biological Supply Co., Burlington, North Carolina 27215. If you prefer, you can have your students collect these organisms themselves. On their own time, such as on weekends or after school, have students locate a nearby pond, lake, stream, or even a ditch which contains water. Have them take along an empty container or jar with a stopper, and ask them to collect a sample of water, surface scum, or bottom ooze, all of which may contain cysts or motile stages of various protozoa. After these samples have been brought to the laboratory, the students can develop cultures by adding grass, leaves, or bits of boiled lettuce to a jar or other suitable container, and adding enough water to more than cover the material. This should then be allowed to stand in a warm place away from direct sunlight.

Frogs, Worms, Insects

These too can either be ordered from a biological supply house or obtained by students. They may be kept in the laboratory in small homemade cages. Suitable cages can be made from large-sized coffee or fruit cans or from wide-mouthed fruit or preserve jars. The tops of these containers may be fitted either with cheesecloth or wire mesh.

B. Preserved Materials

A number of different situations are involved here: First, the sealed jars of display materials. These may be quite small or rather large and require no special care other than cautious handling to prevent breakage of the container or jostling and resultant damage to the contents. A particularly old jar may develop small cracks in the seal, allowing for evaporation of the preservative. It is advisable to check these periodically so that possible cracks can be sealed. Lost fluid can only be replaced by a trained person, so guard against evaporation as much as possible. Students should not tilt display jars since they may leak preserving fluid.

Smaller containers present greater problems because they are handled by students more frequently and may be broken or their contents damaged. If such containers are on display or are to be passed around, you must be certain to instruct students in gentle handling. It is not uncommon to see them shake jars containing delicate specimens such as jellyfish. Be careful!

To prevent damage, small specimens which are to be removed from storage containers for closer observation or dissection, should be removed by the teacher or laboratory assistant with an appropriate tool. Damage usually results if students remove the specimens from their containers. You will have to judge this in relation to the group involved. Once the specimens have been removed from their containers students should carry them to their work areas on a slide, watch glass, dissection pan or tray, or other suitable equipment.

Large preserved materials, such as frogs, fish, pigs, should be removed from their containers with tongs, washed in water to remove preserving fluid, and placed immediately on a dissecting tray. This helps to eliminate some of the irritating fumes. As some individuals are sensitive (skin sensitivity) to preserving fluids, it is advisable to have them use rubber or disposable plastic gloves when handling these specimens. Students should also be reminded not to inhale the fumes of the preservative. Small paper or plastic bags can be supplied for the disposal of the dissected parts of specimens. Again, special arrangements should be made with the janitorial staff for removal and disposal. Containers of preserved materials should be tightly sealed when not in use. Care should be exercised with specimens imbedded in plastic, to avoid surface scratches. Normal care in handling all other specimens should be exercised.

PERTINENT FACTS

- Students should be trained to use care, to be neat, and to clean up after themselves when working in the laboratory.
- The instructions of the teacher should always be followed in order to avoid accidents and dangerous situations.
- Students should always obtain the permission of the teacher before devising and performing experiments.
- Caution should be exercised when handling acids and other corrosive materials.
- Students should be familiar with the location of safety equipment and the procedures to be followed in case of accident.

POSSIBLE QUIZ

1. What are three safety rules to observe when handling chemicals in the laboratory?

2. Explain why acid should never be added to water.
3. What purpose might be served by the addition of grass or boiled lettuce to a sample of pond water?
4. Describe the location and function of *all* safety equipment in your laboratory.
5. Describe a method by which you would dispose of liquid chemicals in the laboratory.

READINGS

Davis, H. M., *Scientific Instruments You Can Make.* Washington, D.C.: Science Service, Inc., 1954.

Lynde, C., *Science Experiences with Ten-Cents Store Equipment.* Princeton, New Jersey: D. Van Nostrand and Co., 1950.

Manual of Laboratory Safety (also a chart on safety). New York: Fischer Scientific.

FILMS

"Basic Principles of the Analytical Balance." 19 minutes, sound, bw, $2.90. United States Public Health Service, Washington 25, D.C.

"Biological Techniques: Measuring Techniques." 14 minutes, sound, color, $5.65. AIBS with Thorne Films, Inc., 1220 University Avenue, Boulder, Colorado.

"Biological Techniques: Weighing Techniques." 8 minutes, sound, color, $3.15. AIBS with Thorne Films, Inc.

"Care of Laboratory Animals." 10 minutes, color, with 33 1/3 r.p.m. record. Filmstrip. United States Public Health Service, Washington 25, D.C.

Lesson 2

THE CARE AND USE OF THE MICROSCOPE

Laboratory time: 90 minutes

AIM

To develop an understanding of the use, care, and operation of the light microscope.

MATERIALS

Microscope; Lens paper; Letter "e" slide; Colored threads slide; Prepared slide of unstained fibers; Clean glass slides; Eye droppers; Beakers of water; Threads of different colors; Salt and sugar crystals; Hair fibers; Nail clippings; Reading glass magnifiers.

PLANNED LESSON

1. Carrying the microscope

The arm of the microscope should be pointed out to your students on a chart or actual microscope. The base should also be identified. After each student is assigned one of the class microscopes, he should be instructed to carry it with one hand grasping the arm and the other placed under the base.

2. Parts of the light microscope

Before using the microscope, it is essential that the student be familiar with its parts and their function. It should be brought out that the light microscope is an instrument used to control light in ways that magnify small objects; that the lenses, which are found in the eyepiece (ocular) and the objectives, are essential components; and that both the eyepiece and the objectives are composed of a system of lenses arranged in such a manner as to produce magnification of a certain order. A duplicated copy of a microscope diagram, as in Figure 2-1, might be given to each student for labeling and future reference. Parts to be pointed out should include:

A. Ocular or eyepiece

With the aid of Figure 2-1 students will be able to see that the ocular is located at the top of the body tube. Magnification power is usually engraved on the ocular itself.

B. Objectives

A revolving nosepiece is located at the lower end of the body tube. Attached to this nosepiece are two or three objectives or lenses, having varying powers of magnification. Generally, these are $10\times$, $44\times$, and $97\times$. It should be mentioned that the $97\times$ lens is known as the *oil immersion lens* and is not commonly used. It should also be remembered that the total magnification of an object would be determined by the combined magnification of the ocular (eyepiece) and objective. *What would be the total magnification when using a $10\times$ ocular and a $30\times$ objective?*

C. Adjustment knobs

Students will be able to see two sets of knobs located at the sides of the microscope. By turning these knobs, they will observe that the larger one causes a more noticeable movement of the body tube than does the smaller. Students should be aware that the large knob is known as the *coarse adjustment* while the smaller one is the *fine adjustment*.

D. Stage

Mention the fact that this platform, located beneath the nosepiece, is used to support slides to be observed, while the central opening allows light to pass through the stage.

E. Substage condenser and iris diaphragm

These structures which serve to control the amount of light should be identified. Students should note that by moving the small lever, the

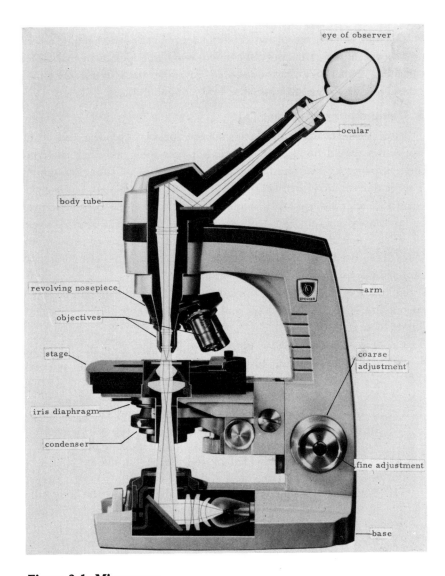

Figure 2-1. Microscope.
Courtesy American Optical Company

aperture of the iris diaphragm can be made larger or smaller, resulting in a corresponding change in the amount of light.

F. Base

This horseshoe-shaped structure upon which the microscope rests should be observed. Although a hinge is present to allow for tilting of the microscope, this is not recommended. *Why?*

G. Light source

It should be mentioned that the light source varies, depending on the microscope model. Generally the microscope will have a mirror which can focus light from an outside source. Although sunlight was once used extensively as a source, lamps are now commonly used.

3. Using the microscope

Before students begin using the microscope, the eyepiece and objective lenses should be cleaned with lens paper only, as other materials such as paper towels, handkerchiefs, and tissue may cause scratching. The ocular should never be removed by students. If it is necessary to clean the inner surface of the ocular, it should be done by the teacher or laboratory assistant. Students should also be warned *never to remove an objective from the microscope,* as the ocular and objectives of each microscope have been carefully coordinated. As a result, parts of microscopes cannot and must not be interchanged.

Skill in focusing the microscope and controlling the light can be learned only through practice. You may find it desirable to have students follow these steps:

- A. After the microscope has been gently placed on the table, the light source should be turned on and the diaphragm adjusted for maximum light entry.
- B. The coarse adjustment knob should *always* be turned toward you so that the nosepiece moves away from the stage.
- C. By exerting gentle pressure, the nosepiece should be revolved until the low power objective (10×) clicks into place.
- D. A prepared slide provided by the teacher should be placed on the stage and fastened with the stage clips.
- E. Watching from the side, the student should lower the body tube, using the coarse adjustment knob, until the objective is *just above but not touching the cover glass.*
- F. Looking through the eyepiece, the student should raise the body tube (the objective moving away from the slide) until it comes into focus. (Turn coarse adjustment *toward* you.)
- G. The iris diaphragm lever should now be manipulated to admit

varying amounts of light. Students should now be able to determine when the proper amount of light is coming through.
H. After it has been observed under low power, the object viewed should be centered, before the student changes to high power. Again, watching from the side, the student should gently revolve the nosepiece so that the high power objective (44×) clicks into position. Students should note that the body tube does not have to be moved, as an object which is in focus under low power will be approximately in focus under high power, and will require only slight manipulation of the fine adjustment knob. *Can high power be used if the objective does not click into place without having to raise the body tube?*
I. After use, the body tube should be raised and the slide removed from the stage. The light source should be turned off and the low power objective placed in position. The body tube should then be lowered and the microscope returned to its proper storage place.

PERTINENT FACTS

- It is not necessary to wear eyeglasses when using the microscope unless they correct for astigmatism. *(Glasses, if used, should also be cleaned.)*
- While using the microscope the student should be instructed to keep both eyes open to avoid eye strain.
- To be viewed clearly, an object must be properly centered under the objective.
- Under high power magnification a smaller portion of the specimen is seen.
- Students must not be allowed to remove the eyepiece or objectives, or to interchange these parts with other microscopes.

4. Using the microscope further

You may want to develop the laboratory further by using slides of various types to illustrate graphically certain characteristics of microscope function. These might include:

A. Letter "e" slide

These can either be purchased or made quite easily. To make the slide, simply provide students with typed or printed letter "e's," a clean glass slide, an eye dropper, and a beaker of water. Students should moisten the clipping and place it on the slide so that the letter is in a normal position when observed with the naked eye. The slide should then be placed on the stage and observed while moving the slide from left to right. Sketch the observed image. *In what direction does it appear to move? In what position does the letter appear to be? Inverted? Reversed? Normal?*

B. Colored Threads Slide

Supply students with three pieces of differently colored threads, a clean glass slide, and a beaker of water. After the threads are wet, they should be placed on the slide so that all three cross at one point. The position of the threads with reference to one another should be noted. Place the slide under the microscope, and observe the point at which the threads cross under low power. *Is it possible to focus on all three threads at the same time? If not, what must be done to get each one in focus?*

C. Fiber Slide

A prepared slide of three unstained fibers may be obtained from a biological supply house. These slides usually contain fibers of cotton, linen, and hemp, and are useful in illustrating the value of proper illumination and the use of the iris diaphragm. Since the fibers are unstained, they are almost invisible when the diaphragm is wide open. As the diaphragm is slowly closed, the finer structure of the fibers becomes visible.

D. Additional Slides

Time permitting, students might be allowed to prepare slides using such materials as crystals of sugar or salt, hair from the eyebrows, nail-clippings, and other readily available materials.

5. Magnifiers

You can bring out the relationship of the microscope to simpler types of magnifiers by having your students write their names on a piece of notebook paper to be observed through a clean glass slide. Have them hold the slide at different positions above the paper. *Does the size of the letters change?* Have them place a drop or two of water on the slide and repeat their observations. *What changes can be observed in the appearance of the letters? How can you account for this?*

Now have them use a reading glass to observe the letters in their name. *In what ways are your observations with the hand lens similar to those through the drop of water?* You can also have them compare the shape of the lens in the reading glass with the shape of the drop of water, and describe any similarities or dissimilarities.

PERTINENT FACTS

- The reading glass, microscope, and drop of water are all types of magnifiers.
- Only extremely thin sections of materials are visible under the microscope since light must be able to pass through them.

POSSIBLE QUIZ

1. What is the meaning of the $10\times$ and $44\times$ engraved on the objectives of the microscope?
2. When a $10\times$ eyepiece is used with a $12\times$ objective, what will the magnification be?
3. What does the term magnification mean?
4. What is the purpose of the fine adjustment?
5. Briefly explain the chief limitations of the ordinary microscope.

READINGS

Beiser, A., *Guide to the Microscope.* New York: E.P. Dutton, and Co., Inc., 1957.

Danielli, J.F. (ed.), *General Cytochemical Methods,* Vols. 1 & 2. New York: Academic Press, 1958, 1961.

Lenhoff, E.S., *Tools of Biology.* New York: The Macmillan Co., 1966.

Wyckoff, R.W.G., *The World of the Electron Microscope.* New Haven, Conn.: Yale University Press, 1958.

FILMS

"Biological Laboratory Techniques: The Microscope." 10 minutes, sound, color, $4.15. McGraw-Hill Book Co., Text-Film Division, 330 W. 42nd Street, New York, N.Y.10036.

"The Compound Microscope." Sound, color, free. Bausch and Lomb Optical Co., Film Distribution Service, 635 St. Paul Street, Rochester 2, New York.

"The Microscope and Its Use." 10 minutes, sound, bw, $2.00. Young America Films, Inc., 18 E. 41st Street, New York, N.Y. 10017.

Lesson 3

THE PREPARATION OF PERMANENT AND TEMPORARY SLIDES

Laboratory time: 90 minutes

AIM

To develop techniques for the preparation of simple slides.

MATERIALS

Clean glass slides; Depression or well slides; Cover slips; Beakers of water; Eye droppers; Petroleum jelly; Scalpel or single edge razor blade; F.A.A. and Boun's fixatives; Varying concentrations of alcohol; Formalin; Glacial acetic acid; Beakers or Coplin jars; Xylol; Paraffin; Paper boats; Slicing apparatus (microtome); Heat source (radiator or light bulb); Stains; Balsam.

PLANNED LESSON

1. Temporary slides

Only a few methods of slide preparation can be included here since there are a large number of special techniques. If you desire further information, consult specialized texts such as those listed in the "Readings" section of this lesson.

A. Simple Wet Mounts

A number of cells and tissues can be mounted in a drop of fluid, usually water. Water is an excellent mounting fluid for living materials. Some difficulties to be considered include the rapid evaporation of water, rapid loss of oxygen from the water, and the entrapment of air under the tissue. Mineral oil may also be used. The visibility of some cell parts is better in mineral oil than in water, and tissues remain alive for longer periods of time. Some materials which can be prepared as wet mounts include protozoa, onion epidermis, elodea leaves, cheek cells, and non-living materials.

The procedure is quite simple: A drop of water is placed on a clean glass slide and the material to be mounted is placed in it. A very small amount of material should be used so that light can pass through. A cover slip should then be placed over the specimen. (It is important that no air be trapped under the cover slip.) The cover slip should be placed on the slide so that one edge is touching the slide, with the remainder of the cover slip held away from the slide, as illustrated in Figure 3-1. The cover slip should then be slowly lowered so that no air is trapped beneath it.

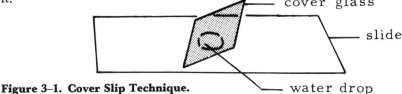

Figure 3–1. Cover Slip Technique.

B. Hanging Drop Slides

Because of the curved surface of a drop of water, light is reflected in many different directions. This problem can be avoided by "flattening" the drop of water with a cover slip, but this also reduces movement when protozoans or other organisms are being studied.

The hanging drop method may be used here. A drop of culture is placed on the center of a cover slip and a depression slide (well slide) is placed over it so that the depression is directly over the drop. The slide and cover slip are then grasped and turned over quickly. When done properly, the drop will hang suspended from the cover slip. Students should practice this technique. For a firm attachment of the cover slip with less likelihood of slippage, the well may be ringed with petroleum jelly before being placed on the cover slip.

C. Stained Temporary Mounts

There are many structures which are not visible unless they have been stained. A number of stains and staining methods may be used to make these parts easy to observe. The most satisfactory are the stains known as *vital stains*. These do not kill organisms immediately as do some others, but are absorbed slowly so that organisms continue to function for some time. Included in this category are Congo red, neutral red, and methylene blue. Specific suggestions for the use of these stains have been included within specific lessons. Those stains which cause the immediate death of cells include Lugol's iodine, Wright's stain, and Gentian violet.

For all types of slides, some problems may arise in slicing techniques. Relatively rigid materials, such as stems and leaves, may easily be cut freehand, with a sharp scalpel, knife, or single edge razor blade. For fresh and preserved tissues, extra care must be taken. If fresh tissue is being cut, the knife blade should be dipped in water before each slice is made. When slicing preserved material, the blade should be dipped in preservative. In each case, the slices should be transferred to a watch glass containing the appropriate fluid.

PERTINENT FACTS

- Water is the most frequently used fluid for the preparation of wet mounts.
- The time spent observing a wet mount is dependent upon the rate of evaporation of the fluid used.
- A very thin section of almost any material can be made into a wet mount.
- Care must be taken to prevent trapping air beneath the cover slip.
- Vital stains are used when it is desired to observe a living specimen and its functions for a prolonged period of time.

2. Permanent slides

The major consideration in preparing permanent slides is to maintain the materials in as natural a state as possible. This requires that cells be killed rapidly. Various materials known as fixatives will perform this function and will also harden the tissue for slicing. There are a number of fixative agents, the more common ones including F.A.A. (formalin, alcohol, acetic acid), which is very good for plant materials; 70% ethyl alcohol, for small organisms and tissue; Bouin's fixative, excellent for both plant and animal tissue; and a number of others which are of specific value for specific types of work. Detailed lists, including instructions for preparation and use, may be found in histology textbooks, such as those given in the list of "Readings" at the end of this lesson.

Formulae:
(a) F.A.A. (formalin, alcohol, acetic acid):
Alcohol, 50% 90cc
Formalin 5cc
Glacial acetic acid 5cc
or
Alcohol, 70% 85cc
Formalin 10cc
Glacial acetic acid 5cc
or
Alcohol, 95% 50cc
Formalin 10cc
Glacial acetic acid 2cc
Water 38cc

F.A.A. can be used to store tissues for long periods of time without damage to them. It has the advantage of not requiring a water bath before dehydration, which is required before staining can begin.

(b) Bouin's fixative:
Formalin 5cc
Glacial acetic acid 1cc
Picric acid 25cc

Tissue should remain in this fixative for 1-2 days before washing with 70% alcohol to remove the color.

A standard procedure is involved in preparing permanent slides. This includes:

1. obtain tissue;
2. transfer to fixative material;
3. pass through a series of alcohols to dehydrate;
4. clear tissue;
5. prepare for sectioning—imbed if necessary;
6. section;
7. use xylol bath to remove paraffin;
8. repeat step 3;
9. stain;
10. counterstain;
11. dehydrate with alcohol;
12. mount on slide.

Materials needed for this procedure include: tissue to be prepared; clean glass slides; varying concentrations of alcohol; xylol; paraffin;

paper boats; slicing apparatus; a source of low grade heat, such as a radiator or light bulb; stains; cover slip; balsam.

Specially designed jars, known as Coplin jars, are usually used for submerging the slide in the appropriate fluids, although small beakers may also be used.

After the materials have been assembled, you will need approximately 30-40 minutes for preparation. The time depends on individual technique and the stains selected. Specific procedures that can be followed include:

1. As thin a section of tissue as possible should be used. Students should practice this.
2. Selection of the fixative will depend on the tissue, although those mentioned previously should meet most of your needs.
3. Dehydration can generally be accomplished by leaving the tissue in 70% alcohol for approximately 6 hours. For delicate tissue, a series of alcohols should be used, beginning with 30% alocohol for 1 hour, 50% for 1 hour, to 70% for 2–3 hours. Finally, the tissue should be placed in absolute alcohol for 1 hour.
4. Xylol is commonly used to remove alcohol, and is known as a clearing agent. Tissues should remain in xylol for at least 2 hours.
5. Use small cardboard containers known as "boats" (which may be obtained from any biological supply house). Fill with paraffin which has been melted in a beaker on a low heat source, such as a radiator or hot plate, or light bulb. Add the tissue to the melted paraffin in the paper boats and allow to "cook" for approximately 3 hours. Cool by placing the beaker in cold water or by removing the beaker from the heat source for 1–2 hours. Remove the hardened paraffin with the tissue imbedded within it, and use the microtome.
6. Section by following the instructions for the use of your microtome.
7. Place a few sections of the sliced tissue on several clean glass slides, and quickly run through a flame once. Do not hold the slide over the flame for any length of time.
8. Place the slide in xylol for 5 minutes.
9. Place the slide in 95% alcohol for 3 minutes; then in 70%, 50%, and 30% alcohol for 2 minutes each; finally in distilled water for 1 minute.
10. Stain as per instructions for the particular stain to be used. (These are far too numerous to list, but can be obtained from the "Readings" at the end of this lesson.)
11. Counterstain.
12. Rinse in absolute alcohol, and transfer to xylol for approximately 30–60 minutes.

13. Mount in a drop of balsam and cover with a cover slip.

PERTINENT FACTS

- The nature of the tissue to be prepared will determine the selection of appropriate materials.
- A series of alcohol baths of varying strengths is necessary to prevent damage to tissue structure.
- The use of proper technique is essential for success.

POSSIBLE QUIZ

1. Fully explain the possible techniques which may be used for preparing a temporary wet mount of protozoa. Explain the advantages and disadvantages of each method.
2. What are the differences between vital and non-vital stains?
3. Why should care be taken to prevent the formation of air bubbles under the cover slip?
4. What are the advantages of wet mounts as compared to permanent slides? The disadvantages?
5. What is the purpose of imbedding tissue in paraffin? Explain the need for using varying concentrations of alcohol.

READINGS

Conn, H.J., *Biological Stains*, 7th ed. Baltimore, Williams and Wilkins Co., 1961.

Conn, H.J., M. Darrow, and V. Emmel, *Staining Procedures Used by the Biological Stain Commission*, 2nd ed. Baltimore: Williams and Wilkins Co., 1960.

Gray, P., *Handbook of Basic Microtechnique*, 3rd ed. New York: McGraw-Hill, Inc., 1964.

———, *The Microtomists Formulary and Guide*. New York: Blakiston (McGraw-Hill), 1954.

Gurr, E., *Encyclopedia of Microscope Stains*. Baltimore: Williams and Wilkins Co., 1960.

Humason, G., *Animal Tissue Techniques*. San Francisco: W.H. Freeman and Co., 1962.

Jensen, W., *Botanical Histochemistry*. San Francisco: W.H. Freeman and Co., 1962.

FILMS

"Biological Laboratory Techniques: Fixing and Section Cutting." 12 minutes, sound, color, $4.40. McGraw-Hill Book Co., Text-Film Division, 330 W. 42nd Street, New York, N.Y. 10036.

"Biological Laboratory Techniques: Staining." 10 minutes, sound, color, $3.90. McGraw-Hill.

Lesson 4

LOOKING AT CELLS AND JUDGING THEIR SIZE

Laboratory time: 45 minutes

AIM
To introduce the principles of microscopic measurement and their use in the observation of cells.

MATERIALS
Microscope; Clean glass slides; Plastic millimeter ruler; Yardstick; Strips of clear plastic; Felt tip pen; Straight pins or dissecting needles; Aluminum foil; Cork cells; Single edge razor; Onion skin cells; Elodea; Dried leaves; Protozoa; Stop watch, if available.

PLANNED LESSON
1. Units of microscopic measurement
Supply your students with a clear plastic millimeter ruler and have them focus upon it using the low power objective of their microscope. If you prefer, you can have the students make their own ruler by supplying them with strips of clear plastic, a felt-tip pen, and a yardstick.

Students should be aware that the lighted circular area which is seen through the microscope is called the *field of vision*. What is the actual diameter *of the field of vision?* Your students can check their estimate of the diameter of the field of vision in the following manner: Supply students with straight pins or dissecting needles and pieces of aluminum foil. Have them make a hole in the foil which approximates what they think is the diameter of the field of vision. This piece of foil should then be placed on a glass slide and observed under low power. *How does this help you relate apparent size, as viewed through the microscope, with actual size?*

In order to measure microscopic objects, a smaller unit of metric measurement than the millimeter must be used. This unit is called a micron, and is equal to 1/1000 of a millimeter. *How many microns are there in a millimeter? What is the diameter in microns of the field of vision observed under low power?*

PERTINENT FACTS
- One micron equals 1/25,000 of an inch.
- One micron equals 1/1,000 of a millimeter.
- One millimeter equals 1000 microns.
- Approximate field diameters of the three standard magnifications are: 100X = 1500 microns; 430X = 350 microns; 970X = 150 microns.

2. Estimating cell sizes

Several types of tissues are particularly suited for introductory experiences with microscopic measurement. Because of the linear arrangement of its cells, cork is especially well suited. Using a single edge razor, students should be instructed to cut a wedge-shaped piece of cork. A small piece should then be cut from the thinnest edge, mounted in a drop of water on a slide, and viewed under low power. Have your students devise a method for estimating the size of a single cell in microns. As your students already know the diameter of their microscope field, you might suggest that they count the cells in a single row that stretches across the field. By dividing the total number of cells in that row into the already determined diameter of the field, individual cell size can be determined.

For additional practice you may have your students measure the size of onion skin cells, elodea cells (common aquarium plant), and cells of dried leaves.

3. Estimating the speed of living specimens

Have your students observe live protozoa (see Lesson 1) in a hanging drop or simple wet mount, under low power. While observing these organisms, students can use a stop watch to calculate the time required for one specimen to move across the entire field or portions of the field. If stop watches are not available, students can work in pairs, one observing the specimen while the other keeps track of the time by using the second hand of his watch. Since the microscope magnifies size, and not time, by converting microns per second to miles per hour, the speed of the specimen can be determined.

PERTINENT FACTS
- In determining the speed of an organism absolute accuracy is essential in timing because of the short distance the animal will travel.
- It is possible to convert microns per second into miles per hour using the conversion values given in the lesson.

POSSIBLE QUIZ
1. Completely describe one method which can be used to determine the size of the microscope field.
2. Why is it best to use a hanging drop preparation in the observation of protozoa?
3. What is the relationship between microns and millimeters?
4. How can you convert microns per second to miles per hour?
5. How can you determine the speed at which living specimens move across the microscope field?
6. What advantages do you see in using mm's and microns instead of inches? Why isn't the United states on the metric system?

READINGS
Carpenter, P., *Microbiology*. Philadelphia: W.B. Saunders Co., 1961.

Lawson, C.A., and R.E. Paulsen, Eds., *Laboratory and Field Studies in Biology*. New York: Holt, Rinehart and Winston, Inc., 1960.

Turtox Service Leaflet #58. *Measuring with the Microscope*. Chicago: General Biological Supply House, 1957.

Weesner, H., *General Zoological Microtechnique*. Baltimore: Williams and Wilkins Co., 1960.

FILMS
"Biological Techniques: Measuring Techniques." 14 minutes, sound, color, $5.65. AIBS with Thorne Films, Inc., 1220 University Avenue, Boulder, Colorado.

Lesson 5

THE DEVELOPMENT OF CONTROLLED EXPERIMENTS

Lesson time: 45 minutes

AIM

To familiarize students with the scientific method.

MATERIALS

Film.

PLANNED LESSON

1. Steps of the scientific method

The teacher should introduce the topic by leading a discussion which might begin with the question, "What are the characteristics of a good experiment?" During the discussion, all the characteristics suggested by your students should be listed on the chalkboard, and then reviewed by the class, so that a final selection of mutually acceptable points can be developed. You can supplement this list so that all the steps of the scientific method are included, as indicated:

A. Problem

It should be clear to your students that an experiment is designed to solve a particular problem. You can illustrate this by using an everyday problem, such as selecting a prom gown or buying a car.

B. Hypothesis

Students should now propose a likely answer to their problem.

C. Collection of Background Material

All known information relevant to the problem should be collected before designing an experiment.

D. Experimentation

Here, actual procedures for problem solving should be developed. It should be stressed that a valid experiment must have a *control*. Both control and experimental groups should be identical in make-up and must be kept under identical conditions as far as possible. Both groups should receive identical treatment in every respect, except for the single *variable* under study. Selection of this variable will depend upon information derived from prior research. After selection of the variable, all techniques which will be needed must be mastered, before actual experimentation may begin. The experiment should be designed to prove or disprove the *hypothesis*.

E. Compilation of Experimental Data

Accurate observation and recording of results should be stressed. The design of charts, graphs, lists, and other materials, is an important aspect of this procedure and should be developed jointly by the teacher and students.

F. Conclusions

After careful collection of data, students evaluate their results and draw logical conclusions. You should point out that these results are valid only within the limits of their experiment—that the results might be inconclusive, and thus show the need for further experimentation. *Will a single experiment serve to solve a problem?*

2. Application of the scientific method

You may wish to reinforce the ideas developed in this lesson by showing a film. Particularly well suited to this lesson is the film "Controlled Experiments," which develops the basic concepts presented in this lesson by showing actual experiments in progress. Students are encouraged to answer questions posed in the film, and time for their answers is allowed.

After the film has been shown, a brief review of the lesson would logically follow.

Now that your students have been familiarized with the development of controlled experiments, you may wish to devise additional laboratories centering around problem-solving and controlled experiments.

PERTINENT FACTS

- The problem should be clearly and simply defined, and must have only one variable.
- It is important that extensive preliminary research be done so that all pertinent data has been reviewed, and to guard against time-wasting duplication of others.
- The experiment should be designed to clearly prove or disprove the hypothesis.
- A valid experiment is one which will produce the same results every time it is repeated.
- The value of the conclusions reached will depend upon the accuracy with which the entire process has been developed and carried out.

POSSIBLE QUIZ

1. Why isn't experimentation the first step in problem solving?
2. Is a hypothesis necessary? Explain fully.
3. What purpose does a control serve? How does one select pertinent data?
4. To what extent is library research valuable in problem solving?
5. Define validity, reliability, and accuracy.

READINGS

Beveridge, W., *The Art of Scientific Investigation.* New York: W.W. Norton and Co., 1957.

Biological Sciences Curriculum Study, *Research Problems in Biology: Investigations for Students*, Series 1–4. Garden City, New York: Anchor Books, Doubleday and Co., 1963, 1965.

Goldstein, P., *How To Do An Experiment.* New York: Harcourt, Brace and World, 1957.

Lenhoff, E.S., *Tools of Biology.* New York: Macmillan Co., 1967.

Wilson, E. B., Jr., *An Introduction to Scientific Research.* New York: McGraw-Hill Book Co., 1952.

FILMS

"Controlled Experiments." 11 minutes, sound, color, $3.65; bw, $2.15. Indiana University, Audio-Visual Center, Bloomington, Indiana 47405.

"Scientific Method." 12 minutes, sound, color, $3.40; bw, $2.15. Encyclopedia Britannica Films, Inc., 1150 Wilmette Avenue, Wilmette, Illinois 60091.

"Scientific Method in Action." 20 minutes, sound, color, $6.65. International Film Bureau, Inc., 57 E. Jackson Blvd., Chicago, Illinois.

Unit II

THE CELL

Lesson 6

THE CELL MEMBRANE AND MOLECULAR MOVEMENT

Lesson time: 90 minutes
Laboratory time: 90 minutes

AIM
To develop clear understandings of the means by which materials pass through cell membranes and the concepts involved.

MATERIALS
Microscope; Clean glass slides; Cover slips, Ammonia or perfume; Potassium permanganate or ink; Ringer's solution: sodium, potassium, calcium chloride, sodium bicarbonate, distilled water; Animal membranes; Cellophane; Collodion liquid; Test tubes of large diameter; Powdered gelatin; Sand; Funnels; Thistle tubes or other glass tubing; Beakers; Rubber stoppers; Eye droppers; Sodium hydroxide; Sodium citrate; Concentrated nitric acid; Sugar solution; Hydrion paper; Human or frog red blood cells; Elodea; Raw carrot or potato; Petroleum jelly; Protective glasses; Laboratory coats.

PLANNED LESSON

1. Structure of the membrane

Basic to the understanding of the movement of materials across a membrane is an understanding of the structure of that membrane. The cell membrane structure may be illustrated using Figures 6-1 and 6—2, either collectively or separately, depending upon the ability of your group and the time available. Figure 6-2 might be used if you wish to provide a more detailed study of the molecular structure of the membrane.

Living material remains alive as long as it is able to maintain a proper balance with the surrounding environment. If the balance is upset the activity of the organism is disturbed, and will cause the organism to show signs of stress, or even die. Your students should know that the membrane of the living cell has the vital role of regulating the passage of materials into and out of the cell. The membrane is normally selective, and as a result, only certain substances can enter or leave the cell. By the selective

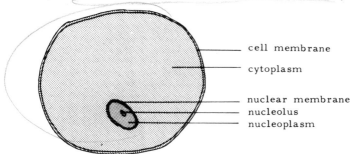

Figure 6-1. Generalized Cell Indicating Membrane and Organelles.

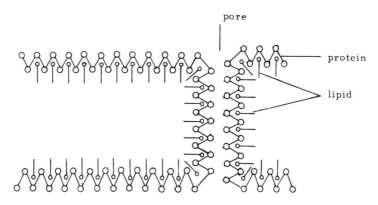

Figure 6-2. Cell Membrane—Molecular Structure.

action of the membrane, the cell can maintain, within certain limits, an equilibrium with the surrounding materials.

2. Diffusion principles and their relation to the movement of various substances.

Using ammonia or perfume (or any substance with a strong odor) you may demonstrate the diffusion of molecules. Leave the bottle uncapped and have students raise their hands as they smell the contents. *Did everyone react at the same time? Why not?*

A crystal of potassium permanganate ($KMNO_4$) or a drop of any colored solution, such as ink, may be placed in the center of a shallow dish filled with water. *What happens? Why?*

PERTINENT FACTS

- Molecules move in random directions.
- Random movement eventually causes even distribution of molecules.
- Molecules move from an area of higher concentration to one of lesser concentration until equally distributed.

3. Diffusion through a membrane—osmosis

With the preceding as an introduction, you can now move on to the movement of molecules through a membrane. To illustrate the selective permeability of the cell membrane, Figures 6-3A, 6-3B, and 6-3C may be used. These figures illustrate suitable set-ups which can be used in a demonstration.

Membranes which may be used include animal membranes, such as sausage skin which can be obtained from most butcher shops, or the stomach and intestine of a frog. To prepare the frog tissue it is necessary to remove the stomach and intestine from a freshly killed frog. The dissected materials then must be washed with Ringer's solution. One end should be tied off with string, while the other end is attached to the apparatus as shown in Figure 6-3. Ringer's solution may be prepared in the following manner:

Sodium chloride (NaCl)	6.000 gm
Potassium chloride (KCl)	0.075 gm
Anhydrous calcium chloride ($CaCl_2$)	0.100 gm
Sodium bicarbonate ($NaHCO_3$)	0.100 gm

Distilled water to make one liter.

The $NaHCO_3$ must be completely dissolved before the $CaCl_2$ is added.

Also suitable to use as membranes are 6-inch circles of cellophane, household plastic wrap, parchment paper, or prepared dialyzing paper. Collodion liquid may be used to prepare collodion bags. The collodion

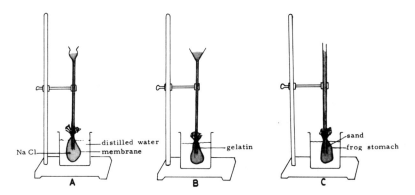

Figure 6-3. Diffusion Apparatus.

should be poured from the stock bottle into a test tube of large diameter. (Rotate the tube while pouring.) Only enough liquid to continuously coat the sides of the tube is needed. Extra collodion may be returned to the stock bottle. After approximately 15 minutes the membrane formed in the tube may be removed. Gently peel back an edge at the top of the tube and then slowly allow running water to flow down between the collodion membrane and the side of the tube. It is sometimes helpful to hang the test tube upside down to hasten drying.

Each of the suggested set-ups should be prepared by filling each membrane with a different type of solution, as follows:

A. Sodium Chloride Solution

Mix 30 grams of table salt (NaCl) with enough distilled water to make a total of 1000 ml (1 liter). The simplest procedure would be to dissolve the salt in a small amount of water in a beaker, pour the solution into a liter flask, and fill with distilled water to the indicator line on the neck of the flask.

B. Colloidal Solution

Add 12-15 grams of powdered gelatin, available in all supermarkets, to about 1 liter of warm water. Add slowly and stir until dissolved.

C. A Suspension

Add 10-12 grams of sand to 1 liter of distilled water, and stir. If possible use only the very powdery parts of the sand.

Quantities may be reduced according to the number of demonstrations.

You should make certain to fill the membranes *before* they are placed in the beaker of distilled water. This will prevent the accidental introduction of solutions into the beaker. For best results, a funnel should be used so that liquids will flow into the glass tubing; or the tube used may actually be a funnel or thistle tube, as Figures 6-3B and 6-3C indicate.

When the membranes have been prepared and filled, they are then gently suspended from a clamp, as illustrated, so that the membrane is beneath the surface of the water in the beaker. (You should allow 1 hour for the preparation of this demonstration. It is advisable to prepare one day in advance so that the materials may be tested.)

After approximately 20 minutes, students can determine which substances moved through the membrane. They should be encouraged to test the beaker solution and observe the movement of water in the tubes. The following tests may be used in testing the contents of the beaker:

A. NaCl

Taste the water. *Salty?* Remove 1–2 ml and place in a test tube. Add a few drops of silver nitrate (test for chloride ions). A white precipitate indicates the presence of chloride ions. *Did sodium chloride pass though the membrane?*

B. COLLOID

Since gelatin is a protein, the test for proteins is applicable here. Remove 1–2 ml of solution from the beaker, place in a test tube, and add 2–3 ml of concentrated nitric acid. If protein is present, a characteristic yellow color will appear.

Care must be exercised when handling acids. Students should wear protective glasses and clothing. If an acid gets on the skin, it should be rinsed off immediately with clear running water. You should prepare a weak solution of a base, such as sodium hydroxide (NaOH), which can be used to neutralize acid by pouring on it. Such a solution should be made in advance and kept in a prominent place known to all students.

C. SUSPENSION

Visual observation should be sufficient to determine the presence of sand particles in the beaker.

The teacher will have to consider time here to determine whether the apparatus will be set up during class or before class begins.

Diffusion in a more natural environment may be demonstrated through the use of a raw potato or carrot. Equipment can be set up as illustrated in Figure 6-4.

Remove a central cylinder leaving approximately $\frac{1}{2}$ inch of the solid portion at the bottom (an apple corer works well here). Slowly pour concentrated sugar solution into the cylinder (molasses may also be used). Close the top with a one-hole rubber stopper through which a piece of glass tubing has been inserted. Place this in a beaker of distilled

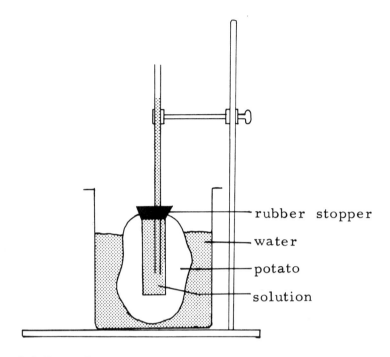

Figure 6-4. Potato Osmometer.

water, as illustrated. Coat the stopper with petroleum jelly to prevent leakage. Water diffuses through the membranes of cells into the tube with some sugar. *How can we determine if sugar molecules diffused out of the central cylinder into the surrounding water?* (You may find it necessary to scrape the surface of the vegetable to improve passage through it.)

Some alternate demonstrations which might be used include:

Prepare an egg osmometer by cutting and removing a small section of shell from the blunt end of an uncooked egg. Be careful not to puncture the underlying membrane. The egg should be placed in a beaker of water, as illustrated in Figure 6-5, so that the water level is just below the broken edge of the shell. The shape or curve of the unbroken membrane should be observed at the beginning of the demonstration, at 20-minute intervals during the class period, and again the following day. *What evidence of a pressure difference has been observed? How can you account for this?*

Using three colored flower petals, students should prepare three different wet mount slides. Mount one petal (or part of a petal) in distilled water, one in a weak acid (dilute hydrochloric acid is good), and the third in a weak base (dilute sodium hydroxide or ammonium hydroxide). Since flower pigments change color with a change in acidic or basic

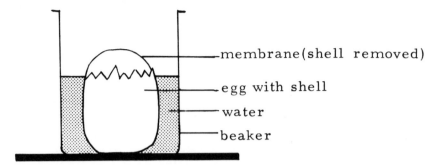

Figure 6–5. Egg Osmometer.

environment, students can observe the effects of these fluids with a microscope. *What do these changes tell you about the movement of the liquids through the cell membrane?*

PERTINENT FACTS

- Molecules of dissolved material (solute) move from an area of higher concentration to one of lower concentration.
- Water moves from an area of greater concentration of water to one of lower concentration of water.
- In osmosis, although many types of molecules move across or through the membrane, the greatest number of molecules are water molecules.
- Thus, osmosis is generally defined in terms of water movement.
- The force with which water molecules move through a membrane is called osmotic pressure.

4. Hypertonic, hypotonic, and isotonic solutions

This demonstration can be used in conjunction with lecture-discussion of the characteristics of these solutions. To demonstrate the effects of hypertonic, hypotonic, and isotonic solutions on living plant and animal cells, red blood cells and elodea (Anacharis) may be used. Appropriate solutions must be prepared in advance ($\frac{1}{2}$ hour required for preparation). Solutions should be prepared as follows:

A. Hypertonic

Add 15 grams of table salt (NaCl) to enough distilled water to make 1000 ml of solution.

B. Isotonic

Add 8.5 grams of NaCl to enough water to make 1000 ml of solution. If pH 5.0 isotonic saline is required, add 0.1N HCl drop by drop and test with Hydrion paper until pH 5.0 is indicated.

C. Hypotonic

Use distilled water.

In order to insure good results, *distilled water must be used as the solvent in each case.* Solutions should be placed in clean, dry eye dropper bottles. There should be a sufficient number of bottles for each table, or pair of students.

Red blood cells may be obtained from any of the following sources:

1. Hospitals—Outdated human blood can usually be obtained free of charge from your local hospital.
2. Animals—From a freshly killed frog or other laboratory animal. (This may coagulate very quickly; be sure to add sodium citrate to frog blood.)
3. Poultry—From a local live poultry market.

Place a drop of blood in the center of a clean glass slide. (Clean slide with 70% alcohol and rub dry.) To prevent clotting, have your students add a drop of 0.9% sodium citrate solution to the blood on the slide and stir gently with a clean toothpick. (To prepare sodium citrate solution, mix 9 grams of sodium citrate into enough distilled water to make a 1 liter solution.) To keep the blood from spilling off the slide, have your students draw a large circle on the center of the slide with a crayon or china marking pencil. Students should then observe the cells under the microscope. After observing normal red blood cells (you may suggest or require a sketch here) students should place a drop of hypertonic NaCl solution on the slide so that it comes in contact with the blood. Have students observe at 5-minute intervals. They should record observations in a sketch. This procedure should be repeated for both isotonic and hypotonic solutions, using a fresh drop of blood and a clean slide in each case.

The entire experiment may be repeated using elodea leaves (common aquarium greens) which can be obtained as the whole living plant from any pet shop. Have your students observe differences in plant and animal cells.

You should encourage your students to use previous knowledge concering diffusion and osmosis to explain observed phenomena. Size and concentration of molecules are two factors which should be considered.

PERTINENT FACTS

- Animals cells placed in a hypertonic solution tend to shrink.
- *Crenation* is the term used to refer to this shrinking of the entire cell.
- When a plant cell is placed in a hypertonic solution, the cell contents will shrink, while the cell wall remains unchanged.

- *Plasmolysis* is the term that refers to this shrinking of the contents of plant cells.
- Animal cells placed in a hypotonic solution tend to swell and burst.
- *Cytolysis* is the general term that refers to the bursting of a cell.
- When a plant cell is placed in a hypotonic solution the cell contents swell, but the cell wall prevents bursting.
- This excess water creates an internal pressure known as turgor pressure (see Lesson 16).
- Isotonic solutions cause no physical changes in plant or animal cells.

POSSIBLE QUIZ

1. When a cell is bathed in distilled water, what molecular movement takes place?
2. What conditions must exist in order that osmosis may occur in a living system?
3. Which of the three types of solutions must be used for intravenous feeding. Why?
4. Why doesn't a protein pass through a semi-permeable membrane?
5. Salt is often added to weeds in driveways to kill them. How does this work? Explain.

READINGS

Adler, B., and T.E. Wainwright, "Molecular Motions," *Scientific American*, October, 1959.

Halter, H., "How Things Get into Cells," *Scientific American*, September, 1961.

Robertson, J.D., "The Membrane of the Living Cell," *Scientific American*, April, 1964.

Solomon, A.K., "Pores in the Cell Membrane," *Scientific American*, December, 1960.

FILMS

"Cell Biology: Transfer of Materials (AIBS, Part I)." 28 minutes, sound, color, $8.15. McGraw-Hill Book Co., Text-Film Division, 330 W. 42nd Street, New York, N.Y. 10036.

"Cells and Their Functions." Sound, bw, $4.50. Athena Films, Inc., 165 W. 46th Street, New York, N.Y. 10019.

"Osmosis." 16 minutes, sound, color, $6.65; bw, $4.15. Encyclopedia Britannica Films, Inc., 1150 Wilmette Avenue, Wilmette, Illinois 60091.

Lesson 7

GENERALIZED CELL: INTERNAL STRUCTURE AND FUNCTION

Lesson time: 45 minutes

Laboratory time: 30 minutes

AIM

To acquaint students with the structure and functioning of cell parts, stressing their interrelationships and developing the understanding that the cell is the basis of all life.

MATERIALS

Overhead projector; Microscopes; Transparency; Clean glass slides; Cover slips; Stains—Janus Green B, Methylene blue, Lugol's iodine, Gentian violet; Celery; Onion cells; Cheek cells; Straight edge razor; Toothpicks; Eye droppers; Sugar solution; Beakers of water.

PLANNED LESSON

1. Cytoplasm

Using an overhead projector and a transparency of a generalized cell throughout (a particularly good diagram of a generalized cell can be obtained from "The Living Cell," by Jean Brachet, p. 5, in a special issue of selections from *Scientific American*, September, 1961), the teacher might quickly review the characteristics of colloids (Lesson 6), pointing out that the cell protoplasm is made up of water (bound H_2O) attached to protein molecules (which are quite large). This produces a colloid. *What are some of the characteristics we would expect the cytoplasm to show? Since the unattached or free water is needed for the chemical reactions of the cell, what conclusions can be drawn concerning cell activity in terms of amount of free water?*

2. Mitochondrion

The teacher has much leeway here concerning depth of presentation. In most cases it would be advisable to simply identify this "powerhouse of the cell" as the structure in which enzymes cause the production of energy by breaking down glucose—the ultimate results being the production of water and carbon dioxide. You should point out that these results strongly indicate that mitochondria are the actual sites of cellular respiration.

If greater depth is desired, you can use Figure 7-1 to indicate the *cristae* and the location of enzymes within a mitochondrion. A brief introduction of the energy storage molecules, ATP, should also be included. These molecules pick up energy at the surface of a mitochondrion and store it for future use. It should also be pointed out that mitochondria are usually found in areas of the cell where much energy is needed. These areas include the *synapse* of nerve cells, *contractile fibers* of muscle cells, at the base of *cilia*, and the *endoplasmic reticulum*, where energy is needed for protein synthesis.

The following has proven to be a very useful and interesting student laboratory exercise to observe enzyme activity of mitochondria: Janus Green B (which can be obtained from the chemistry room of your school, from a local hospital, or a local medical supplier) or methylene blue is used to stain the mitochondria of cells in celery, and then the dye is bleached by the various enzymes produced by these structures.

Cut a thin slice from a fresh celery stalk. Transfer this 1 mm-thick, half-moon shaped slice to a clean glass slide so that the freshly cut surface faces you. Add a drop of 5% sugar solution. A series of dots, now visible, are the strings (*collenchyma*). Using a sharp instrument, such as a single edge razor blade, cut away the celery tissue so that only a center section between two strings remains.

Place a cover slip over the cells on the slide, and look for cytoplasmic movement or streaming (*cyclosis*). Also find the green *plastids* (ovoid bodies along the sides of the cell), clear nucleus, and small moving spheres and rods which are the mitochondria.

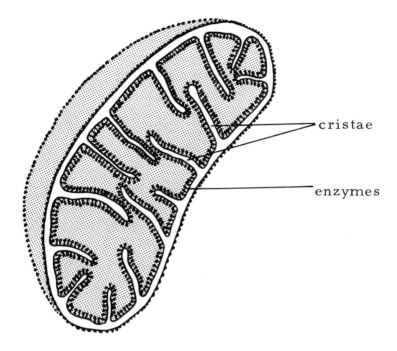

Figure 7-1. Mitochondrion.

Apply filter paper to one end of the cover slip, while at the same time adding a few drops of a .001% solution of Janus Green B to the opposite side of the cover slip. Through the microscope students will be able to see the mitochondria being stained blue; then within minutes, they will be bleached or decolorized by enzymes (*dehydrogenases*), on the mitochondria.

3. Lysosomes

"The digestive tract" of the cell, the lysosome is a *vacuole* which contains digestive enzymes, envelops particles of food, foreign material, and worn out cell parts to be digested through the process of *phagocytosis*. Students should be able to explain *why the digestive enzymes do not leak through the lysosome membrane.*

4. Endoplasmic reticulum and ribosomes

(Only seen in Electron microscope photomicrographs.) The site of protein formation, the reticulum is a series of membranes running through the cytoplasm, the outer surface of which is covered with small round bodies called ribosomes. The membranes increase the surface area of the cell, while the ribosomes synthesize various cellular proteins.

5. Golgi apparatus

Mention should be made to your students that this apparatus appears to be continuous with the reticulum, and seems to have some sort of secretory function within the cell. It is generally believed that the Golgi apparatus plays a role in the packaging of certain proteins as secretory granules of the cell. The shape of this apparatus varies but it generally can be identified near the nucleus of the cell. In the plant cell, it is thought that this apparatus is involved in the formation of the cell plate.

6. Vacuoles

Clear, bubble-like structures can be seen within the cytoplasm of most plant and animal cells. These are surrounded by a double membrane similar to all other membranes in the cell. Students should be aware that these vacuoles perform a number of functions within the cell, usually relating to storage or excretion. Digestive enzymes may be added to vacuoles, resulting in the production of lysosomes.

7. Cell differences

Apart from the structures presented thus far, which appear to be part of most cells, you should point out basic similarities and differences between generalized plant and generalized animal cells. For example, you may point to the presence of *centrosomes* and *centrioles* in animal cells, and their absence in plant cells; the presence of *chloroplasts* or green plastids in plant cells, and their absence in animal cells.

Some simple laboratory activities which may be used to illustrate cell differences include:

To study plant cells, supply your class with fresh onions. The teacher should slice the onion into thick sections, each approximately $\frac{1}{2}$ inch thick. Students then separate the concentric rings of a slice and carefully peel away a piece of the transparent "onion skin" found between rings. A simple staining technique can be used to prepare the tissue for observation. (Lugol's iodine is suitable as is Gentian violet.) The tissue is carefully spread out on a clean slide using a probe or toothpick. A drop or two of stain is placed on the tissue and allowed to remain for 1–2 minutes, after which the stain is carefully rinsed off by using an eye dropper, gently squeezing water on the slanted slide. To prevent stain from running onto tables or sinks, have the students place a paper towel underneath the slide. A cover slip should be placed on the slide, excess liquid carefully removed, and observations made.

Animal cells may be studied by preparing a slide of cheek cells. Each student should use a clean toothpick from a freshly opened box. By gently scraping the inside surface of the cheek and spreading this material across a clean slide, the cheek cells will be ready for staining. Repeat the process used for staining onion cells.

PERTINENT FACTS

- As new techniques and equipment are developed and invented, more facts about the internal structure of the cell are brought to light.
- The mitochondria have now been identified as the sites of respiration in the cell.
- Vacuoles become lysosomes when the cytoplasm adds enzymes.
- Lysosomes break down not only food particles but also foreign living material and worn out cell parts.
- Recent investigations have shown the presence of Golgi apparatus in plant cells; prior to this, it was thought that these organelles were present only in animal cells.

POSSIBLE QUIZ

1. Specify three kinds of cytoplasmic bodies and tell how each is visually identified.
2. What are some structural differences between plant and animal cells?
3. What might the function of cyclosis or cytoplasmic streaming be in the cell?
4. How are the various cytoplasmic bodies functionally interrelated?
5. Differentiate between cytoplasm and protoplasm. Is there a difference between the two? Explain fully.

READINGS

Brachet, J., "The Living Cell," special issue of *Scientific American*, September, 1961.

Brachet, J., and A.E. Mirsky (eds.), *The Cell*. New York: Academic Press, 1959-1961.

Gabriel, M., and S. Fogel(eds.), *Great Experiments In Biology*. Englewood Cliffs, New Jersey: Prentice-Hall, Inc., 1955.

Lehninger, A., "Energy Transformation in the Cell," *Scientific American*, May, 1960.

Loewy, A.G., and P. Siekevitz, *Cell Structure and Function*. New York: Holt, Rinehart and Winston, 1963.

Swanson, C.P., *The Cell*. Englewood Cliffs, New Jersey: Prentice-Hall, Inc., 1964.

White, P., *The Cultivation of Plant and Animal Cells*. New York: Ronald Press, 1954.

FILMS

"Cell Electron Micrograph Overhead Transparency." Set TOT-30, $60.00 per set of 12. General Biological Supply House, Inc., 8200 South Hoyne Avenue, Chicago, Illinois 60620.

"Cells and Their Function." Sound, $4.50. Athena Films, Inc., 165 W. 46 Street, New York, N.Y. 10019.

"Protoplasm, The Beginning of Life." Sound, $3.00. Bray Studios, 729 7th Avenue, New York, N.Y. 10019.

MODELS

1. Students can use candy, clay, or plastic, to make either a whole cell or a single cell structure.
2. Commercial plastic models of cells are available at many hobby stores.

Lesson 8

THE GENETIC CODE

Lesson time: 45 minutes

AIM
To investigate current materials dealing with the nucleus and chromosomes, including a modern approach to DNA and RNA.

MATERIALS
Plastic ladder such as found in bird cages; Felt board; Colored paper; Film.

PLANNED LESSON

1. Nuclear membrane

The teacher should emphasize its structural similarity to and continuity with all other membranes of the cell, and that it encloses the portion of the cytoplasm known as *nucleoplasm*.

2. Nucleolus

This structure within the nucleus serves as a temporary storage place for RNA. Students should be aware that there may be several nucleoli present within a single cell.

3. Chromosomes

Early observations by microscopists showed that rod-shaped structures known as chromosomes were present in the nucleus of cells. Experimentation indicated that sections of these chromosomes, later named genes, were the structures that carried hereditary traits.

In 1953, Dr. Watson and Dr. Crick proposed a three-dimensional model of chromosome structure. They hypothesized that a chromosome was in fact a molecule of *deoxyribonucleic acid* (DNA).

Briefly the teacher can present DNA as being "ladder-like" in structure, and show its spiral arrangment using the following *demonstration*:

Obtain a plastic ladder, such as those sold in pet shops and used in bird cages. By holding each end of the ladder and twisting it in opposite directions, a spirally coiled structure will be achieved.

If you wish to explain DNA structure in more detail, several options are available:

A. The Plastic Ladder

A. *The plastic ladder*, used above, may be prepared in advance by using pieces of colored paper attached to the sides of the ladder to represent the sugar and phosphate molecules of the side chains. Strips of paper may be placed on the rungs to show the attachment or bonding of bases. A different color should be used for each of the four bases.

B. A Felt Board

B. *A felt board* may be prepared, either by the teacher or by your students, using Figure 8-1 as a pattern. Any color combination may be used, although it is suggested that related bases be coordinated in color. Suggested colors might be: phosphate-blue; sugar-green; adenine-yellow; thymine-orange; guanine-red; cytosine-pink. If the teacher prefers, this may also be used as a chalkboard diagram with colored chalk.

Students can be encouraged to construct models of DNA using materials of their choice. They should be aware that they can predict the bases which will be found on the "right hand side" of the ladder, if the bases of the "left hand side" are known. When this concept has been established, the teacher may then move on to the reading of the genetic code.

4. Genes

Genes, or segments of the chromosomes, control specific characteristics. With the aid of the teacher, students should develop an understanding of the Watson-Crick interpretation of a gene. Using the DNA models previously developed, the concept that a series of bases arranged in a sequential pattern are the segments of chromosomes which are called genes, can be developed. These genes control all metabolic functioning of the cell.

THE GENETIC CODE

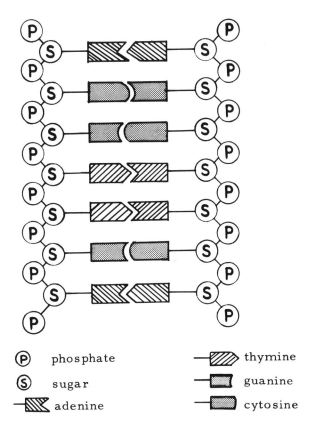

Figure 8-1. DNA Structure Model.

5. RNA

Molecules similar to DNA which are stored in the nucleolus, carry the code from the nucleus to other parts of the cell. Known as *ribonucleic acid* (RNA), it is believed that these molecules are structurally similar to DNA. The teacher may wish to point out basic differences between the two—a different sugar in the side chains; and a base known as *uracil* in RNA, in place of the base *thymine* in DNA.

Three types of RNA should be mentioned: *ribosomal* RNA (rRNA), which is believed to be a structural component of ribosomes; messenger RNA (mRNA), which serves to carry the genetic code from the genes to the ribosomes on the endoplasmic reticulum; and, transfer RNA (tRNA), which transfers amino acids from the cytoplasm to the ribosomes during the process of protein synthesis. Transfer RNA is sometimes referred to as soluble RNA, with the symbol sRNA.

6. Extending the lesson

An important part of this lesson should be the showing of a film, such as "The Thread of Life," produced by the Bell Telephone Company. This one hour color film may be obtained free of charge by contacting your local Bell Telephone office and may be used during a regularly scheduled laboratory period.

It should be pointed out that Lessons 8 and 10 can be combined, as time permits, at the discretion of the teacher.

PERTINENT FACTS

- The nucleus is separated from the cytoplasm by the nuclear membrane.
- Within the nucleoplasm is found the nucleolus and chromosomes.
- It is now believed that a chromosome is a DNA molecule.
- The nucleus, containing the genes, therefore controls all cellular activity.
- Several types of RNA have been identified and are involved in protein synthesis.

POSSIBLE QUIZ

1. With the aid of diagrams, briefly discuss the structure of a DNA molecule.
2. In what ways is DNA similar to RNA? How do they differ?
3. Discuss the role of DNA and RNA in relation to protein synthesis in the cell. What is the function of the nucleus in the cell?
4. Discuss the various types of RNA and their functions.
5. Discuss the gene in relation to the Watson-Crick model of DNA.

READINGS

Clark, B.F.C., and K.A. Marcker, "How Proteins Start," *Scientific American*, January, 1968.

Hanawalt, P.C., and R.H. Haynes, "Repair of DNA," *Scientific American*, Feburary, 1967.

Herman, D.J., and Li W. Laphan, "Genetic Passenger," *Scientific American*, February, 1968.

Horowitz, N.H., "The Gene," *Scientific American*, April, 1956.

Marley, F., "Interlocked DNA; Molecules of DNA From Mitochondria," *Scientific American*, January, 1968.

Mirsky, A.E., "The Discovery of DNA," *Scientific American*, June, 1968.

Swanson, C.P., *The Cell*. Englewood Cliffs, New Jersey: Prentice-Hall, Inc., 1964.

Zamekof, et. al., "Biologically Active DNA Synthesized at Sanford," *Science Teacher*, March, 1968.

FILMS

"DNA, The Molecule of Heredity." 16 minutes, sound, color, Encyclopedia Britannica Films, Inc., 1150 Wilmette Avenue, Wilmette, Illinois 60091.

"Genes and Chromosomes (AIBS)." 28 minutes, sound, color, $8.15. McGraw-Hill Book Co., Text-Film Division, 330 W. 42nd Street, New York, N.Y. 10036.

"The Thread of Life." 1 hour, sound, color, free. Contact your local Bell Telephone office or write to: American Telephone and Telegraph Co., Motion Picture Section, 195 Broadway, New York, N.Y. 10007.

MODELS

1. Prepared DNA models can be obtained from the following sources:
  A. DNA Model Kit, based on the Watson-Crick theory, Burgess Publishing Co., 426 S. 6th Street, Minneapolis, Minnesota 55415. An interested student preparing this model should be ready to explain and demonstrate it in class.
  B. DNA Kit for Student Use, composed of plastic pieces. KD Biographics, 1050 Flake Drive, Palatine, Illinois 60067.
  C. Students may develop models of their own using whatever materials they have on hand, such as plastic, candy, or clay.
2. An alternative to the preceeding might be to prepare a model based on the story in the October 4, 1963 issue of *Life Magazine* (this has been updated and is now in reprint form). Life Educational Reprint Program, Box 834, Radio City Post Office, N.Y., N.Y. 10019.

Lesson 9

CELL DIVISION

Lesson time: 90 minutes
Laboratory time: 90 minutes

AIM

To develop the concepts of mitosis and meiosis, stressing the functions of each.

MATERIALS

Microscopes; Glass slides; Cover slips; Prepared slides of mitosis; Fresh onions; Glass jars or beakers; Toothpicks; Acetocarmine stain: glacial acetic acid, distilled water, powdered carmine; Watch glass (pyrex); Films.

PLANNED LESSON

1. Why do cells divide?

Most cells undergo regular division. Have students suggest possible explanations concerning the necessity for such division. Develop the idea that division is triggered by an increase in the amount of cytoplasm to the point where the nucleus can no longer control it efficiently, and a sufficient quantity of material cannot be brought into the cell for normal functioning. Advanced classes might consider the relationship of the area of a sphere (cell membrane) to the volume of a sphere (cytoplasm). It will be seen that the membrane grows at a rate in the order of r^2 (area of a sphere-πr^2) while the cytoplasm increases in the order of r^3 (volume of a sphere-$4/3\pi r^3$).

Point out to your class that there are two common forms of cell division, mitosis and meiosis. Mitosis, which results in a quantitative production of cells (an increase in the number of cells), is responsible for growth. Mitosis is the process by which all *somatic* cells reproduce. Cells thus produced are identical to the parent cell. Meiosis results in the production of germ (reproductive) cells, which have half the number of chromosomes of the parent cell.

2. Mitosis

Remind your students that mitosis occurs in living cells with the exception of reproductive cells (*gametes*). An outline of the stages of mitosis and a simple sketch for each stage, similar to Figure 9-1, should be developed on the chalkboard or overhead projector as the lecture-discussion progresses.

A. Interphase

It should be pointed out to your students that the interval which begins immediately after *telophase* and continues until division occurs again is called *interphase*. (The appearance of a cell in interphase would be identical with the sketch of the generalized cell, since in fact, they are the same.) During interphase all cellular functions except division occur. It is therefore not advisable to use the older term "resting stage." During interphase, the *chromatid* or single strand of DNA must undergo the process of *replication*. During replication the single strand of DNA will be transformed into its double helical structure. In other words, the chromatids become chromosomes. Until replication has been completed, the nucleus cannot begin a new mitotic division.

B. Prophase

Chromosomes become visible and appear as long threads in the prophase. This is caused by the contraction of the DNA molecule. Each chromosome appears as two identical halves which are attached at a single point. Students should know that these halves are known as chromatids and are joined at the point known as the *centromere* (*kinetochore*). During this stage the nucleoli and nuclear membrane begin to disappear. The end of prophase is marked by the total disappearance of these structures as well as a cessation of chromosome contraction.

A new structure, known as the "spindle," begins to appear as the nuclear membrane disappears. It should be pointed out that this process

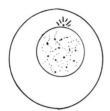
interphase

early prophase

mid-prophase

early metaphase

late metaphase

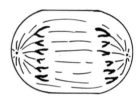
anaphase

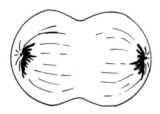
early telophase

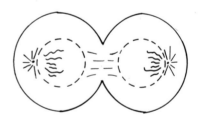
late telophase

daughter cells

9–1. Stages of Mitosis.

differs in plant and animal cells. In animal cells the fibers which make up the spindle radiate from structures known as centrioles, and are formed from a single structure, the *centrosome*. They migrate along the nuclear membrane until they are opposite each other. After this migration, spindle fibers begin to radiate from them. In plant cells, centrioles are not present, although a functional center, or *pole*, does exist.

C. Metaphase

During this stage, chromosomes move toward the center of the cell, known as the *equatorial plate*. Chromatids remain attached to each other. Spindle development continues until fibers stretch from pole to pole. Chromosomes become attached to the fibers at the centromeres. By the conclusion of metaphase, all chromosomes will be aligned at the equatorial plate.

D. Anaphase

During *anaphase* chromatids move toward the poles. This movement is made possible by the division of the centromere, which allows for separation of each chromatid pair. Students should be aware that the mechanism for chromatid movement has not been explained, although the spindle fibers are thought to play a prominent role. Anaphase ends when chromatids form a densely packed group at the poles. Each group will have the same number of chromatids in it. Since each chromosome divided into two chromatids, each of which moved to one pole, it can be shown that each group of chromatids must contain a number of chromatids equal to the number of original chromosomes.

E. Telophase

During the *telophase*, the nuclear membrane and nucleoli reappear while the chromatids become less visible. In animal cells, the centrosome reappears as the spindle disappears. You should stress that mitosis is the process by which two new nuclei are formed; therefore, it ends with the formation of the new nuclear membranes.

Generally, *cytokinesis* (the division of the cytoplasm), occurs simultaneously with nuclear membrane formation. The teacher should stress that mitosis always results in an equal division of chromosomal material, while cytokinesis does *not* always result in an equal division of cytoplasm.

It should further be emphasized that there are differences in the process as it occurs in plant and animal cells. In the plant, a *cell plate* begins to form across the equatorial plate of the cell. This will divide the cytoplasm and will give rise to the cell wall. In animal cells, the cell membrane begins to pinch in at the center in a process known as "furrowing." This pinching-in will ultimately divide the cytoplasm in half.

As a consolidation of the material presented, the film "Mitosis" might well be used.

3. Observation of cells undergoing mitosis

Select several fresh onions, and suspend them in small glass jars or beakers by inserting three toothpicks into the side of each onion at its center. The toothpicks should be placed so that they suspend the onion in an upright position. Fill the jar with water so that the base of the onion is covered. The onions should be kept out of direct sunlight for several days, after which roots will have developed.

There are a number of alternate methods for the preparation of the roots. The simplest method would be to heat slices of the root tips in acetocarmine dye. Acetocarmine can be prepared by:

    Glacial acetic acid . . . . 45 ml
    Distilled water . . . . 55 ml
    Powdered carmine to saturate

Dissolve enough powdered carmine dye in the acetic acid until it becomes saturated. Add the distilled water to this, and bring the solution to a boil. Make certain that the room is well ventilated. After the solution has cooled, filter. Add two drops of ferric chloride solution before using. As acetocarmine is unstable, it should be freshly prepared for each use.

A small Pyrex watch glass should be used for heating the roots. (They should *not* boil.) Four or five minutes should suffice. It will be observed that the root has not been uniformly stained. Only the darkest section should be used. This segment should be cut away and transferred to a clean glass slide. Students should be instructed to mash this segment with the head of a pin or other instrument. After covering with a cover slip, the material may be mashed further by gently pressing on the cover glass with a finger. (The cover glass should be covered with a piece of paper to prevent finger-print smears on the cover slip.) *It is important that this mashing be done thoroughly so that one or two cell layers will be produced.*

Students should be able to observe various stages of mitosis on their slides by using both low and high power. They should also observe their classmates' slides so that they can see all stages. The teacher might make a set of commercially prepared slides available for comparison. He should also remind his students that in living tissue many cells will be undergoing division and thus they will have to study an individual cell to determine the stage in which it is found. (This exercise will require 90 minutes. The stain needs to be prepared beforehand, and will require 20 minutes for preparation.)

PERTINENT FACTS

- Cell division seems to be triggered by an increase in volume of cytoplasm, making normal cell functioning difficult.
- There are two major types of nuclear division, mitosis and meiosis.
- Although generally included in a description of telophase, the division of cytoplasm is actually a separate process called cytokinesis.
- Although basically the same, cell division in plant cells differs from cell division in animal cells in several respects; the main differences include the presence of the centrosome in animal cells, and the way in which the cytoplasm is divided.

4. Meiosis

Stressing the differences between mitosis and meiosis once again, the teacher should encourage students to consider the special function of meiosis. They must understand that each type (species) of animal has a characteristic, fixed chromosome number, referred to as the *diploid number* of chromosomes. They should consider the process of sexual reproduction and the effect of the union of sperm cell and egg cell on the total number of chromosomes. *How can a continuous doubling of chromosome number be avoided?* They should learn that some specific process is needed to reduce the number of chromosomes. You can develop the idea that a special kind of cell division causes a reduction of the number of chromosomes in the gametes (reproductive cells). This is known as the *haploid number* of chromosomes, and is the result of reduction division or meiosis.

Meiosis occurs in two parts—meiosis I and meiosis II—each of which has four steps, as does mitosis. The teacher will find that a review of mitosis will prove valuable at this point. After refreshing the basic ideas of mitosis, you should present the stages of meiosis in lecture-discussion by outlining the stages on the chalkboard or overhead projector, using illustrations similar to those in Figure 9-2.

MEIOSIS I:

A. Prophase

Students should be aware that this stage is the same as the prophase stage of mitosis.

B. Metaphase

Homologous pairs of chromosomes line up at the equatorial plate. It is now known that all cells have pairs of chromosomes which are identical in appearance and are referred to as being homologous. (One pair,

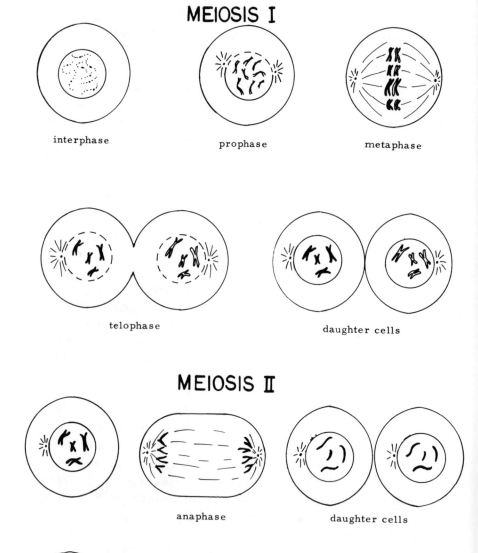

Figure 9-2. Stages of Meiosis.

the sex chromosomes, are not identical.) This is a significant difference from mitosis, where chromosomes are aligned haphazardly. All other conditions are the same as in mitosis.

C. Anaphase

A significant difference, as compared to mitosis, occurs here: the centromere *does not* divide. This means that chromatids cannot move to the poles, but rather *whole chromosomes* (pairs of chromatids) move. One member of each homologous pair of chromosomes goes to each pole. This stage causes the "reduction" of chromosome number.

D. Telophase

Continues as in mitosis. Cytokinesis occurs as in mitosis.

MEIOSIS II:

Each new daughter cell, produced in Meiosis I, undergoes a regular mitotic division immediately after Meiosis I. Since this division *always* follows immediately, the *total* process is called meiosis. The effect of meiosis, then, is the reduction of the species number (diploid number) to the half (haploid) number. In the human, the diploid number of 46 would be reduced to the haploid number of 23.

Although the presentation of the concepts of meiosis is frequently looked upon as difficult by many teachers, by stressing concepts developed for mitosis the teacher should have few problems. It should be stressed that there are actually only a few variations in the two processes; if taught from this point of view, many pitfalls can be avoided. Obviously, both processes have been simplified. The teacher may add more detail as the ability of the group indicates. It is suggested that all students be thoroughly familiar with these basic ideas before going into greater depth. Students should be reminded that meiosis occurs in plants, as in the formation of spores which give rise to haploid gametes, as well as in the formation of animal gametes.

As a consolidation of the material presented, the film "Meiosis" might well be used.

PERTINENT FACTS

- Meiosis is a cell division which occurs only in the formation of gametes, in both plants and animals.
- Meiosis encompasses two nuclear divisions, but only one reduction of chromosome number.
- The main differences between mitosis and meiosis occur in Meiosis I when homologous pairs of chromosomes are aligned at the equa-

torial plate (metaphase), and when the centromere does not divide (anaphase), and so chromatids do not separate.
- Meiosis is essential to the maintenance of a fixed species number of chromosomes.
- The species number of chromosomes is also known as the diploid number, while the half number, found in gametes, is known as the haploid number.

It should be noted that as a continuation of this study, the teacher may follow this lesson with Lesson 27, Unit 4, Reproduction and Development, which follows the development of the fertilized egg to the multicellular organism.

POSSIBLE QUIZ

1. What is the function of mitosis? Meiosis? Briefly outline the major occurrences in each step of mitosis.
2. List the steps of meiosis and indicate only the differences between mitosis and meiosis and the stages in which these variations occur.
3. How many daughter cells will result when one parent cell undergoes meiosis? Explain.
4. Indicate the differences in mitosis as it occurs in a plant cell, and as it occurs in an animal cell.
5. Divide the following into two groups. Group 1—haploid cells, Group 2—diploid cells: spores of higher plants, hair cells, lung cells, egg cells, fertilized egg cells, sperm, leaf epidermis cells, root cells.

READINGS

Alston, R.E., *Cellular Continuity and Development*. Glenview, Illinois: Scott, Foresman and Co., 1967.

Kerr, N.S., *Principles of Development*. Dubuque, Iowa: Wm. C Brown and Co., 1967.

Mazia, D., "How Cells Divide," *Scientific American*, September, 1961.

———, "Mitosis and the Physiology of Cell Division," in J. Brachet and A.E. Mirsky (eds.), *The Cell*, Vol. III. New York: Academic Press, 1961.

Swanson, C.P., *The Cell*, 2nd ed. Englewood Cliffs, New Jersey: Prentice-Hall, Inc., 1964.

FILMS

"Cell Biology: Cell Reproduction (AIBS, Part I)." 39 minutes, sound, color, $8.50. McGraw-Hill Book Co., Text-Film Division, 330 W. 42nd Street, New York, N.Y. 10036.

"Meiosis: Sex Cell Formation." 16 minutes. Encyclopedia Britannica Films, Inc., 1150 Wilmette Avenue, Wilmette, Illinois 60091.

"Mitosis." 24 minutes. Encyclopedia Britannica Films, Inc.

"Mitosis and Meiosis." 16 minutes. Indiana University Films, Audio-Visual Center, Bloomington, Indiana 47405; also NET Film Service, 10 Columbus Circle, New York, N.Y. 10019.

MODELS

1. Clay models may be constructed illustrating the chromosome arrangements of the various stages of mitosis and/or meiosis. Many other materials are also suitable, such as colored pipe cleaners or strips of colored paper.
2. Three-dimensional charts may be prepared by students.
3. Commercially prepared models and charts of plant and animal mitosis, and models of meiosis, can be obtained from: General Biological Supply House, Inc., 8200 South Hoyne Avenue, Chicago, Illinois, 60620.

Lesson 10

HEREDITY

Lesson time: 135 minutes
Laboratory time: 45 minutes

AIM

To develop the basic concepts of the transmission of hereditary characteristics from generation to generation.

MATERIALS

Slide projector; Colored slides illustrating Mendelian genetics; Film; Genetic corn; Seeds of hybrid corn; Seeds of hybrid peas; P.T.C. paper: phenylthiocarbamide, distilled water, filter paper.

PLANNED LESSON

1. Chromosomes: the carriers of genetic information.

As developed in Lesson 8 (The Genetic Code), a chromosome is actually a single molecule of deoxyribonucleic acid (DNA). Building on your students' knowledge of the Watson-Crick model of DNA, you can begin to explain the theory as it relates to coding. (Here, some general information will be needed if the teacher chose to omit Lesson 8 options for deeper study of DNA.) It should be made clear that the arrangement of bases along the molecule can have many variations. Students should be encouraged to consider possible consequences of these variations.

Since it has long been accepted that certain areas of a chromosome known as genes controlled various hereditary traits, it can now be seen that these genes are probably segments of this series of bases. Investigators believe that reading down the DNA molecule three bases at a time, provides the code. These three base groups are known as triplets or *codons*.

2. Genes

It can now be seen that genes are simply a series of codons. *How can we determine the "beginning" or "end" of a message (gene)?* It is believed by scientists that certain codons act as punctuation marks.

3. Heredity in the cell

All cell processes depend upon the information carried by the DNA. *How would this information control the production of proteins at the endoplasmic reticulum?* It can be developed that all enzymes (proteins) which affect and control all the chemical reactions of the cell, are manufactured according to information on the DNA. In this manner, the nature and functioning of the cell will depend upon the information carried by DNA.

4. Heredity within the total organism

Students should expand the concepts just developed so that they may now develop a more familiar understanding of heredity, the heredity of eye color, height, and all the other obvious characteristics of man.

Some basic information should be given your students here so that they will understand the concepts that are involved. You should remind your class of the homologous chromosomes of meiosis. They can consider that the fertilized egg or *zygote*, from which they developed, was composed of 23 chromosomes from one parent and 23 from the other parent. This results in the individual having 22 homologous pairs and one pair of sex chromosomes.

A very simple introductory exercise can be performed, using these particular chromosomes as examples, to indicate the determination of sex. The teacher should explain that two X chromosomes will be found in a female body cell, and one X and one Y chromosome in a male body cell. Draw a circle on the chalkboard with 22 homologous pairs of chromosomes indicated inside, as well as 2 X chromosomes. This should be labeled as a female cell. Use the symbol for female, ♀. A second circle, indicating 22 homologous pairs of chromosomes as well as 1 X and 1 Y chromosome, should be drawn on the chalkboard. This male cell should be labeled with the symbol for male, ♂. (It might be pointed out that the male organism produces sperm which are motile. The arrow on the symbol may be considered to indicate motility.) Using Figure 10-1 as a guide, the teacher should now indicate the results of a complete meiosis division for each of the cells.

After the diagrams have been completed, all the possible unions of sperm and egg should be listed. (Only the sex chromosomes should be indicated, as only sex determination is being considered at this time.) For example, the union of gamete 1-Female with 1-Male would result in a zygote containing two X chromosomes. Thus, 1-Female, 1-Male = XX. A complete list of possible unions is indicated in Figure 10-1.

After this concept has been developed, it should be pointed out that since the male gamete contains both an X and Y chromosome, determination of the sex of the zygote would be dependent upon the type (X or Y) of sex chromosome carried by the fertilizing sperm cell.

5. Using the Punnett Square

Although valuable and useful as illustrations, diagrams are not essential for the successful determination of the results of gametic combinations. The Punnett Square, as illustrated in Figure 10-2, presents a simpler method. Your students might enjoy practice in working with Punnett Squares.

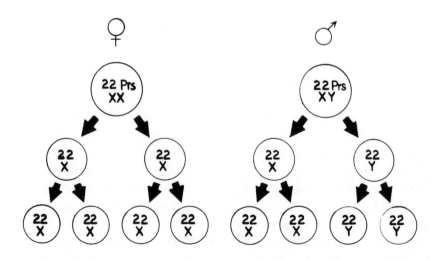

Figure 10–1. Distribution of X and Y Chromosomes.

Heredity

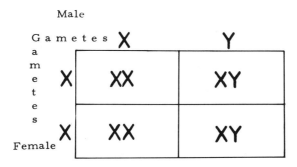

Figure 10-2. XX, XY Cross.

6. Mendelian genetics

The work of Gregor Mendel should be introduced here. If time permits, student reports might be assigned to provide background material. Reference might be made to the film "The Thread of Life" (Lesson 8) if used earlier, or it might be used here. Before introducing Mendel's Laws, you should familiarize your students with the following terms:

Phenotype. The appearance of an organism—what it looks like.

Genotype. The genes an organism contains.

Homozygous (Pure). An organism possessing identical genes for a particular characteristic.

Heterozygous (Hybrid). An organism possessing different genes for a particular characteristic.

Some of Mendel's Laws which should be considered include:

A. Law of Dominance

Some genes are referred to as *dominant*, while others are *recessive*. These terms imply that an organism which inherits one dominant and one recessive gene for the same characteristic will have the physical characteristics determined by the dominant gene. This organism would be said to be heterozygous or hybrid. This implies that the physical manifestation of a recessive gene can occur only when the zygote contains all recessive genes for that characteristic.

The Law of Dominance might be illustrated with a discussion of eye color. Students should be made aware of the dominance of the gene for brown eye color, and the recessive nature of the gene for blue eye color. This would mean that a blue-eyed person must have only recessive genes for blue eye color. A brown-eyed person may have only dominant brown eye color genes, or a combination of both dominant and recessive genes, as illustrated in Figure 10-3.

B. Law of Segregation

When two hybrids are crossed, recessive genes whose characteristics were hidden within the hybrids (because of combination with dominant genes) may be rearranged during the union of gametes, so that numbers of the second generation (F_2) contain only recessive genes for that characteristic. Using a Punnett Square as illustrated in Figure 10-4, the teacher should show the presence of the homozygous recessive.

C. Law of Independent Assortment

To this point, the teacher has presented Mendelian Laws dealing with the inheritance of a single characteristic only. The Law of Independent Assortment, on the other hand, deals with the inheritance of several traits in combination. The characteristics of shape and color of pea seeds, as illustrated in Figure 10-5, are most useful in demonstrating this law.

As the Punnett Square indicates, the *dihybrid* cross results in some offspring which resemble the parents, as well as other offspring whose physical characteristics are different from either parent (i.e., green-round or yellow-wrinkled.) It should be stressed that this is the result of the *random* combination of genes.

D. Law of Incomplete Dominance

Students should learn that, in some isolated cases, the crossing of two pure (homozygous) organisms will result in offspring whose physical appearance exhibits a blending (mixing) of the dominant and recessive characteristics of *both* parents. An example of this can be seen in the "Four O'Clock" flowers. When homozygous white and homozygous red flowered plants are crossed, the offspring will be pink. *What might be expected as the result of crossing two pink Four O'Clocks?*

Color slides can be used with great effect in the presentation of these laws. Commercially prepared slides are available from most supply houses. If these are not available to the teacher, he may prepare such slides with a minimum of difficulty. Standard 35 mm color slides can be used. By photographing various arrangements of flowers such as carnations, a series of slides can be produced demonstrating phenotypes and the results of various crossings between homozygous and heterozygous individuals.

HEREDITY

	B	B
b	Bb	Bb
b	Bb	Bb

Figure 10-3. Dominance.

	B	b
B	BB	Bb
b	Bb	bb

Figure 10-4. Segregation.

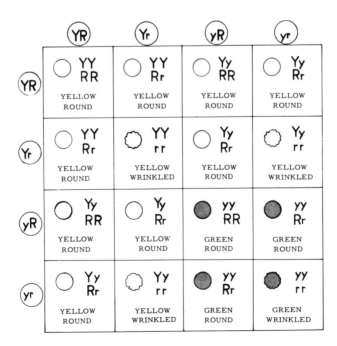

Figure 10-5. Chart of Mendelian Pea Crosses.

In place of these slides, the teacher may use the film "Genetics: Mendel's Segregation." If time permits, "Genetics: Mendel's Recombination" should follow.

PERTINENT FACTS

- When a dominant and a recessive gene are present (for a particular trait) in an organism, the phenotype will be an expression of the dominant character.
- When two hybrids are crossed, recessive genes which previously were hidden may appear phenotypically in the offspring.
- When the inheritance of several characteristics is considered, offspring are produced from a random combination of genes.
- There are cases in which there is a blending of the dominant and recessive genes in the offspring.

7. Laboratory experiences

A number of interesting and worthwhile laboratory activities provide simple and interesting ways for students to apply what they have learned.

A. PHENOTYPES IN CORN

Working in groups of two, with an ear of corn supplied by the teacher, your students should identify as many differences in the appearance of the kernels as possible. Have them make a list of each different type of kernel so that students who have not worked with that group will be able to accurately select a kernel which exhibits those characteristics. A count of the total number of kernels of each type should be made, and that number indicated next to the appropriate description. Students should now be able to determine the proportions of the various types of kernels. This is known as the *phenotypic ratio*.

B. INHERITANCE OF A SINGLE CHARACTERISTIC

The ability of a plant to produce chlorophyll is dependent upon the presence of a dominant gene. Plants which are homozygous recessive for this trait lack this ability. Biological supply houses offer hybrid seeds of corn which are suitable for this activity. Pairs of students should be supervised in the planting of five corn seeds in suitable containers, planted at a depth of one inch. Each container should be clearly labeled with the students' names, section, and date of planting. This may be done at the end of a regular class session and should require approximately 15 minutes. Allow two weeks for development, during which time plants are watered regularly. It will then be possible to count the number of green plants and the number of "albino's."

Each pair of students should determine the ratio of green to albino for their plants. The ratio for all plants can be determined, providing for greater accuracy. The expected ratio is three green plants to one albino (3:1). *Knowing that the genotypes of the parents were hybrid green, is it possible to construct a Punnett Square? If possible, do so.*

C. Phenotypes in Man

Many hereditary traits in humans can be readily studied by observation. Students will enjoy analyzing phenotypes of classmates, and later, phenotypes of family members. Traits to be studied include:

1) *P.T.C. tasters or nontasters.* Phenylthiocarbamide is tasted by approximately 7 out of every 10 people. P.T.C. paper needed for this test may be obtained from a commercial supplier, or it may be prepared by dissolving 500 mg of phenylthiocarbamide in 1 liter of water (requires 24 hours). Sheets of filter paper should be soaked in this solution and hung to dry. When dry, small strips can be cut, and are ready to use. Have all students taste the paper, at the same time. Record the number of tasters and nontasters. The ability to taste P.T.C. paper is a dominant trait (T).

2) *Attached or unattached earlobes.* Have your students observe their classmates and record the number who have earlobes which are attached to the side of the head, and the number of those who have unattached or free earlobes. Unattached earlobes are dominant (E).

3) *Tongue rollers or nonrollers.* Observe and record the number of students who can roll their tongues in a U-shape. This is dominant (U). Folding the tip of the tongue back is recessive.

All of the following traits should be observed and recorded in a similar manner:

4) *Cleft chin.* Cleft chin is a dominant characteristic; smooth chin is recessive.

5) *Dimples.* Dimples are a dominant trait.

6) *Widow's peak.* When the hairline forms a point dipping downward at the center of the forehead, the individual is said to have a widow's peak. This is a dominant trait.

7) *Freckles.* The production of freckles is dominant over the inability to produce freckles.

No doubt students will want to continue their study outside of the classroom. Using members of the family, they can collect data and propose possible genotypes on the basis of their observations. They can then develop a family pedigree from these traits.

8. Genetic aberrations in man

Although the human being has been used as an example in several cases, it might be interesting to pay special attention to various genetic "errors" which may occur in man.

A. Mutations

What is a mutation? This term is used often enough in everyday life so as to be familiar to most students. Discussion should lead to the understanding that a mutation is an alteration in a gene which may lead to the development of new and different characteristics. By far the greatest percentage of all mutations are detrimental mutations. A number of factors may be responsible for the production of mutations. These include radiation (as from excessive use of X-rays or atomic explosions), and drugs (as the breakage of chromosomes caused by LSD). For the further study of the effects of drugs, see Lesson 31, Unit 6. It should be emphasized that the process of mutation is an integral part of evolution.

B. Nondisjunction

Students should be aware that nondisjunction is a case in which, for reasons as yet unclear, chromosomes fail to separate during meiosis I. This results in the production of a daughter cell which contains one *extra* chromosome while the other daughter cell has one *less* chromosome. When the sex chromosomes are considered, three types of abnormal zygotes may result from nondisjunction. These would include the zygote with one less chromosome than normal. This zygote (human) would contain 44 body chromosomes and only one sex chromosome (either X or Y). A zygote containing 44 body chromosomes and the Y chromosome dies soon after formation. Students should be aware that the presence of a Y chromosome in any cell is responsible for "maleness." Therefore, a cell which contains only X chromosomes is female. As a result, cells exhibiting the "nullo-X" (XO) pattern (those which contain a single X and no Y), must be female. The reproductive organs of individuals with the XO pattern are absent or underdeveloped. This condition is known as Turner's Syndrome.

Individuals with the XXY pattern are male, as the Y chromosome is present. Parts of the reproductive systems of these individuals are abnormal. A higher rate of mental retardation has been observed in this group. The condition is known as Klinefelter's Syndrome.

It should be pointed out that in the third group are individuals with the XXX pattern. This is the least common of the three. Mental retardation and sterility are symptoms observed in these females.

HEREDITY

PERTINENT FACTS

- Mutations are changes in genes.
- Most mutations are detrimental to the organism.
- Mutations are an essential part of evolution.
- In nondisjunction, chromosomes fail to separate.
- There are three abnormalities which may result from the failure of the sex chromosomes to separate.

POSSIBLE QUIZ

1. Listed below are hypothetical genotypes. Indicate the kinds of gametes which can be produced. TT; Tt; TTBB; TtBb; ttbb.
2. Indicate the phenotypes of the F_1 generation resulting from a cross between a homozygous tall pea plant and a homozygous short plant. What Mendelian Law does this illustrate?
3. Indicate the genotypes and phenotypes of the F_1 generation of a cross between a tall, round seeded plant and a short, wrinkle seeded plant. Plant tallness and round seeds are dominant characteristics.
4. Show the results of a cross between a dihybrid tall plant with green seeds (TtGg) and a short plant with yellow seeds (ttgg).
5. Recalling that Four O'Clock flowers exhibit incomplete dominance, indicate the results of the following crosses: red x white; pink x pink; pink x white; red x pink; white x white.

READINGS

Auerbach, C., *Genetics in the Atomic Age.* New York: Oxford University Press, 1965.

McCafferty, R.C., and G.R. Easterling, "Understanding Heredity," *School Science and Math*, January, 1967.

Moore, J.A., *Heredity and Development.* New York: Oxford University Press, 1963.

Peters, J.A. (ed.), *Classic Papers in Genetics.* Englewood Cliffs, New Jersey: Prentice-Hall, Inc., 1959.

Sonneborn, T.M. (ed.), *The Control of Human Heredity and Evolution.* New York: The Macmillan Co., 1965.

Wallace, B., *Chromosomes, Giant Molecules, and Evolution.* New York: W.W. Norton and Co., 1968.

FILMS

"Genetic Investigations." 12 minutes, sound, color, $3.40. Indiana University, Audio-Visual Center, Bloomington, Indiana 47405.

"Genetics: Inheritance in Man (AIBS, Part VI)." 26 minutes, sound, color, $8.15. McGraw-Hill Book Co., Text-Film Division, 330 W. 42nd Street, New York, N.Y. 10036.

"Genetics: Mendel's Laws." 14 minutes, sound, color, $5.65. Coronet Films, Coronet Building, 65 E. South Water Street, Chicago, Illinois 60601.

"Genetics: Mutations" (AIBS, Part VI)." 30 minutes, sound, color, $8.15. McGraw-Hill Book Co., Text-Film Division.

"Genetics: Mendel's Segregation (AIBS, Part VI)." 25 minutes, sound, color, $8.15. McGraw-Hill Book Co., Text-Film Division.

"Genetics: Mendel's Recombination (AIBS, Part VI)." 28 minutes, sound, color, $8.15. McGraw-Hill Book Co., Text-Film Division.

"Heredity: Fact and Fallacy." 29 minutes, sound, bw, $5.40. Indiana University, Audio-Visual Center.

MODELS

1. Students can prepare models using clay, plastic, or other readily available materials, showing the arrangement of the chromosomes during the various stages of mitosis and meiosis.
2. Commercially prepared models of mitosis and meiosis can be obtained from General Biological Supply House, Inc., 8200 South Hoyne Avenue, Chicago, Illinois 60620.
3. Basic genetic sets, charts, display mounts, plants, seeds for study, genetic corn, all can be obtained from General Biological Supply House, Inc.
4. Genetic materials, including micro slides, plastomount, and P.T.C. papers can be obtained from Carolina Biological Supply Co., Burlington, North Carolina 27215.

Unit III

PLANTS: FORM AND FUNCTION

Lesson 11

THE ROOT AND ITS FUNCTIONS

Lesson time: 45 minutes
Laboratory time: 70 minutes

AIM

To familiarize the students with basic root structures, and to develop the principles of conduction in roots, including root pressure, cell sap, and gravitational effects.

MATERIALS

Microscope; Glass slides; Cover slips; Petri dishes; Beakers; Glass and rubber tubing; Carrot; Clump of grass; Rooted leaf or stem; Begonia plant; Tomato plant; Geranium plant; Seeds of radish, corn, bean, wheat, grass, cabbage, barley; Sharp knife; Sand; China marking pencil; Dark blotting paper; Oil; Petroleum jelly.

PLANNED LESSON

1. General functions of roots

Before going into an examination of the different parts of a root, a brief discussion might be included concerning some general root functions. Your students should be aware that roots play a role in plant anchorage and support, erosion prevention (*how do roots do this?*), food storage, and absorption and conduction of water.

2. Types of roots

Most roots fall into three general categories. The teacher can show his class examples of each type. The *tap root* can be demonstrated by using a carrot (with greens attached); the *diffuse* with a clump of grass (roots intact); and a rooted leaf or stem cutting for *adventitious* roots. It should be noted that the entire edible portion of the carrot constitutes the so-called tap root. *What possible explanation can you give to account for the thickness of a tap root?* Close examination of the tap root (carrot) should reveal numerous hair-like outgrowths. *What function do these serve?*

After the class has satisfactorily identified the food-storing function of the tap root and the water-absorbing functions of its *secondary* roots, the teacher should go on to diffuse (*fibrous*) roots. Using a clump of grass as an example, the class can readily see that all the roots of this system appear to be of uniform size and that they branch (diffuse) in many directions. *Why is this type of root useful in combating soil erosion?*

Using a leaf or stem cutting from which roots have begun to grow, the adventitious root system can be illustrated. Working with the derivation of the term "adventitious," the class should understand that these roots develop from locations which would not normally produce them. The simplest method for preparing demonstration materials would be to make stem cuttings from a household or class begonia plant. Use a sharp knife to cut pieces at least 3–5 inches long, and place these in a beaker or glass of clean water. After several days, adventitious roots will be clearly visible.

3. Anatomy of roots

Several days before lecture-discussion of root anatomy, the teacher should prepare radish seedlings. Prepare petri dishes or other flat glass dishes by covering the bottom of each with a piece of blotting paper. A dark color paper will facilitate observation of the light colored seedlings. Place 6–12 radish seeds on the blotting paper in each dish, making certain that they are evenly spaced. Add enough water to soak the blotting paper, and cover. The dishes should be placed in a warm, dark area to facilitate germination. During lecture-discussion, students can observe the germinating seedlings as the teacher goes into a discussion of root structure:

A. Root Cap

Several layers of cells at the very tip of the root act to protect it from injury as growth causes it to be pushed through the soil.

B. Meristematic Region

This area of active cell division is found immediately behind the root cap. Most commercial slides of plant mitosis are prepared from this meristematic tissue. This region of cell production is primarily responsible for lengthening of the root.

C. Region of Elongation

Above the meristematic region, the region of elongation contains recently formed cells which are becoming elongated.

D. Region of Maturation

After pointing out that this is the region of cell differentiation (specialization), you might wish to go into a more detailed study of roots, using as an example the root of a *dicot* such as the buttercup (*Ranunculus*). When viewed in cross section at the region of maturation, as in Figure 11-1, the root is seen to be composed of three distinct parts: a central region, the *stele*; an intermediate cylinder of tissues known as the *cortex*; and an external layer of tissue known as the *epidermis*.

1) Stele

The stele is composed of vascular tissues known as *xylem* and *phloem*. The xylem, usually arranged in an "X" pattern, serves to conduct water and inorganic salts from the roots upward, into the stem, and may also conduct food which has previously been stored in the roots. Xylem cells are the chief strengthening tissues of roots. Phloem tissue is located in the angles between the xylem "X." Phloem cells carry manufactured food from the leaves through the stem to the roots.

Parenchyma tissue sometimes separates the xylem from the phloem. Surrounding all of these tissues is the *pericycle*, which is composed of small parenchyma cells, capable of producing new cells which grow out to form branch roots. The central region thus limited by the pericycle is the stele or *vascular cylinder*.

2) Cortex

Composed of large, thin-walled parenchyma cells. The cortex makes up a major portion of the young root, and stores much reserve food, usually in the form of starch. It also serves as a pathway for the movement of water and minerals from the root hairs to the vascular tissues of the stele. A single layer of small rectangular cells, known as the *endodermis*, separates the cortex and the stele.

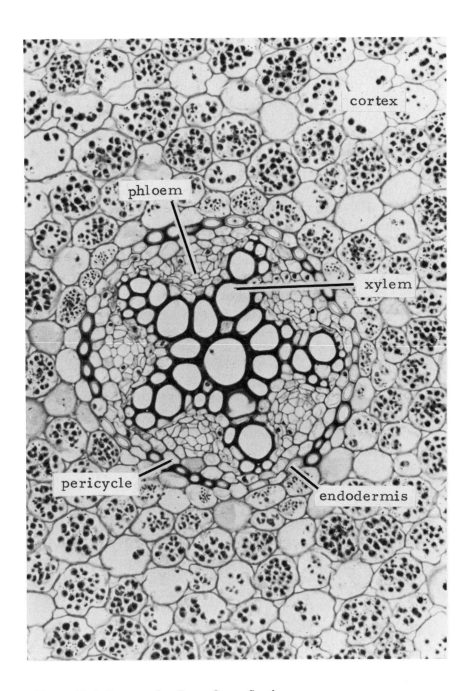

Figure 11–1. Ranunculus Root, Cross Section.
Courtesy Carolina Biological Supply Company

3) Epidermis

This outermost layer of the root is composed of a single layer of cells. The primary function of the epidermis is absorption. This layer produces the root hairs which increase the rate of absorption of the root.

How many of the described root parts can you observe without the aid of a microscope? After your students have become familiar with the parts of the root in lecture, the same seedlings can be used for more detailed study in the laboratory, using the microscope. Have your students prepare a wet mount of a radish root and observe. Utilizing what they have learned in lecture, students should be able to identify the root cap, meristematic region, region of elongation, region of maturation (root hair region). They should neatly sketch their observations and label as many structures as possible.

Additional activities may be included as time permits. Growth in a very young root can be observed by lining a glass jar or beaker with dark blotting paper and filling the container with sand or soil. Soak corn grains for two days in water (soaking reduces the time needed for germination), and then place them between the blotting paper and the glass (the sand will hold them firmly in place), allowing approximately one inch on either side of each grain. Moisten the sand and leave the container in a warm place for several days. The region of maturation can be identified by the root hairs. The glass may be marked with a china marking pencil to indicate the position of the margin of the lower edge of the region. This marking procedure should be repeated at regular intervals not more than a week apart. Data should be recorded at the conclusion of observations, and a graph might be made by your students, plotting change of position in millimeters and time in days.

Other seeds which are suitable for macroscopic observation of roots are bean seeds, wheat, grass, cabbage, barley, rye, and oats. Germination procedures similar to those used for radish and corn seeds are suitable.

PERTINENT FACTS

- The carrot is an example of an edible root.
- The vascular tissues of the root are the xylem and phloem.
- The major regions of a root are the root cap, meristematic region, region of elongation, and region of maturation.
- The meristematic root region becomes differentiated into all other tissues.

4. Physiology of roots

Although not fully understood even by plant physiologists, water absorption may be presented to your students in a general manner. The teacher may briefly review the principles of osmosis (Lesson 6, Unit 2). It is known that osmosis is involved in the intake of water from the soil and in the transfer of water from cell to cell. Also important is the process of *imbibition*, in which the water is absorbed by colloidal particles which then swell. It is known that cells expend energy in active water absorption, but this process is poorly understood as yet. Whether osmosis or imbibition is more important is as yet uncertain.

Solutes are taken into the root by diffusion and active transport, although little is known about the precise mechanism. The combined force of the absorptive powers of the cells, and the movement of water by osmotic pressure, causes the water to move toward the stele. This is a measurable force known as *root pressure*. This root pressure is also responsible for driving fluids up the xylem. It can be seen, then, that a healthy plant will continually have columns of water within it, constantly being replenished from the bottom (roots) as it is lost or used at the top (leaves). *What will happen to a plant which is not adequately watered? How does this relate to the begonia cuttings used earlier in this lesson?*

Have students observe the pruning of a plant in class. They should easily be able to identify the fluid which is exuded and the reason for this "bleeding."

A laboratory activity for the observation of root pressure may be prepared by assembling apparatus as illustrated in Figure 11-2. Use a growing geranium plant. Cut the top of the plant so that a short section of stem remains above the soil (2–3 inches). Attach a piece of glass tubing to the stem by using a short piece of rubber tubing. The glass tubing should be approximately three feet long, and will require support as illustrated. Pour a small amount of water into the tube, so that the water is visible above the tubing. In order to prevent water loss, the rubber tubing may have to be sealed with petroleum jelly. A few drops of oil should be floated on the water in the tube. (Use a funnel to put water water in the tube. Add the oil to this water *before* filling.) Use a china marking pencil to mark the tube, indicating the water level.

Place the apparatus in a warm place and observe. (Be certain that the soil is moist.) After 15 minutes, the water level should be seen to rise as a result of the action of root pressure. Mark the new level. Observe

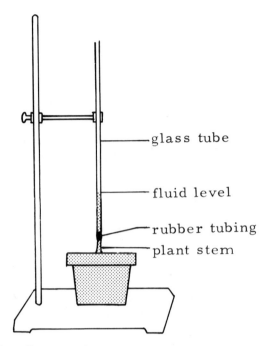

Figure 11-2. Root Pressure Apparatus.

at two-day intervals. Data should be collected. Interested students might plot the result, as was done for root growth. This may be done in conjunction with other laboratory procedures or during lecture. Preparation will require approximately 10 minutes.

Another labratory activity that may be performed to illustrate root pressure involves the use of a tomato plant. Obtain a tomato plant with the leaves intact (they can be raised in flower pots in the classroom, obtained from a nearby greenhouse, or from a neighboring garden). Cut the stem at the base so that a piece approximately 12 inches long is left. Select a piece of rubber tubing which will fit tightly over the cut end and will be long enough to reach a nearby water faucet. Make certain that the ends of the tubing are firmly attached to the stem and to the faucet. *Slowly* turn on the water, and gradually increase the pressure. After a few minutes, droplets of water will be seen to form along the edges of the leaves. This demonstrates the process known as *guttation*. Guttation normally occurs when there is abundant moisture in the soil and atmosphere, causing an increase in root pressure. This rise in pressure forces droplets of water to appear along the leaf edges.

5. Economic importance of roots

Students will be able to give the more obvious values of roots, as in food, beverages, medicines, and prevention of soil erosion. They should be reminded that economic importance also includes *harmful* aspects, such as the invasion of drains, the elevation and breakage of sidewalks, and the toxic effects of some roots on plant and animal life.

The teacher might mention some foods which are derived from roots, such as licorice and marshmallow, as well as the more familiar carrot, parsnip, radish, and sweet potato. Other useful substances would include horseradish, sarsaparilla, and sassafras. It should also be mentioned that some dyes are obtained from roots.

PERTINENT FACTS

- Absorption and conduction are the primary functions of roots.
- The main factors involved in absorption are osmosis and active transport.
- Root pressure is responsible for the movement of sap in plants.
- Guttation is the result of root pressure.

POSSIBLE QUIZ

1. Describe some functions of roots and illustrate with specific examples. How are roots protected as they grow through the soil?
2. What are the differences between root hairs and branch roots?
3. Why would diffuse roots be more effective in controlling erosion than tap roots?
4. What is the advantage of xylem and phloem being close together? Explain how water reaches the vascular tissues from the soil.
5. What is root pressure? How can it be demonstrated?

READINGS

Carlquist, S., *Comparative Plant Anatomy*. New York: Holt, Rinehart and Winston, 1961.

Doyle, W.T., *Nonvascular Plants: Form and Function*. Belmont, California: Wadsworth Publishing Co., Inc., 1965.

Esau, K., *Anatomy of Seed Plants*. New York: John Wiley and Sons, Inc., 1960.

Salisbury, F.B., and R.V. Parke, *Vascular Plants: Form and Function*. Belmont, California: Wadsworth Publishing Co., Inc., 1965.

Steward, F.C., *About Plants: Topics in Plant Biology*. Reading, Mass.: Addison-Wesley Publishing Co., Inc., 1966.

FILMS

"Root Development." 9 minutes, sound, bw, $1.90. United World Films, Inc., 221 Park Avenue South, New York, N.Y. 10003.

"Roots of Plants." 11 minutes, sound, bw, $3.40. Encyclopedia Britannica Films, Inc., 1150 Wilmette Avenue, Wilmette, Illinois 60091.

MODELS

1. Students can prepare a model of a root tip using different colors of clay or wood or other readily available materials.
2. Prepared microscope slides, root tip model, and plastomounts can be obtained from Carolina Biological Supply Co., Burlington, North Carolina 27215, and from General Biological Supply House, Inc., 8200 South Hoyne Avenue, Chicago, Illinois 60620.

Lesson 12

THE STEM AND THE CONDUCTION OF FLUIDS

Lesson time: 45 minutes

AIM

To familiarize students with basic stem anatomy and to further illustrate the principles of conduction and translocation.

MATERIALS

Preserved stems; Celery petiole; Beaker of water; Red ink; Single edge razor blade.

PLANNED LESSON

1. General functions of stems

Before going into an examination of the different parts of a stem, a brief discussion might be included concerning some general stem functions. The teacher should elicit from his students the fact that stems play a role in the production and support of flowers and leaves, as well as in the conduction of materials. Students should be aware that conduction includes movement upward, downward, and transversely. In young (immature) plants, the stem is also involved in food production, but to a small extent. An interesting point to bring out is the fact that some stems, such as the potato tuber, kohlrabi, and onion, serve as storage regions for food. In some *epiphytes* (plants which are supported by poles, wires, or other plants, such as orchids), the stem absorbs water and minerals, taking on the functions of the roots.

2. Types of stems

Elicit from your students the fact that there are two main types of stems—*woody* and *herbaceous*. Woody stems are found in *perennial* plants, or those which live for more than two years. They have large amounts of vascular tissues and cork. Examples of woody stems would be trees, hedges, and certain shrubs.

Herbaceous stems are generally soft and green, and contain no woody tissue. As a result, they do not increase in diameter to any degree. The amount of vascular tissue they form is relatively small, and there is a frequent lack of cork development. As a rule, herbaceous stems are found in *annual* plants, or those whose life span is only one season. Most garden plants are herbaceous. Students should be encouraged to observe various plants, and to identify the stem type.

3. External anatomy of stems

Preserved stems may be obtained from commercial supply houses and used in conjunction with the lecture-laboratory period. Living specimens, such as geraniums, may also be used. Figure 12-1, illustrating external stem anatomy, will be most helpful in identifying external stem features.

Students should be able to identify buds, *lenticels, nodes, internodes,* bud and twig scars, and bundle scars. *How can you obtain information to help in this identification?* The teacher will have to make previous arrangements with the librarian to borrow books for classroom use as well as to have students use the library during the class period.

4. Internal anatomy of stems

As a continuation of the material developed in Lesson 11, Roots, students might be asked to enumerate possible similarities between stems and roots. It should be apparent that vascular tissues will extend through the stem. In addition to xylem and phloem, the cortex and epidermis also are present. In order to reinforce the concept that there is a continuity from root to stem, the three main structural components studied in Lesson 11 should be reviewed here, with emphasis on modifications as they appear in the stem:

A. Stele

As illustrated in Figure 12-2 on internal stem anatomy, the stele in the stem contains vascular bundles and pericycle, as does the root. A difference which will be noted is the presence of a central layer of thin walled parenchyma cells, known as the *pith*, which serves as a storage area.

A second difference is the presence of a layer of meristematic tissue known as the *cambium*, which separates xylem from phloem. It will be recalled that in the root, parenchyma tissue served this purpose. The cambium is responsible for the production of secondary xylem and phloem, the xylem being produced at the inner edge of the cambium (nearer the pith), while the phloem is produced at the outer edge. In cross section it can be seen that secondary phloem comprises a thin layer of tissue located at the perimeter of the stem, while the xylem comprises several rings of tissue interior to the phloem. Students should be aware that seasonal changes affect the deposition of vascular tissue. As in the root, the pericycle lies external to the vascular bundles. *How can the age of a tree be determined through observation of a stem cross-section?*

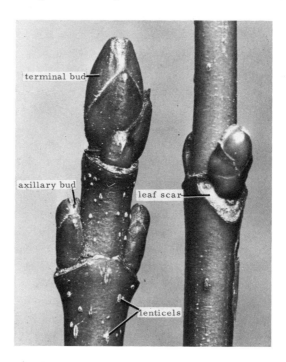

Figure 12-1. External Stem Anatomy.
Courtesy Carolina Biological Supply Company

B. Cortex

As in the root, the cortex is a layer of large, thin-walled cells surrounding the stele. These cells include *collenchyma* and fiber cells, in addition to parenchyma cells which are also found in the cortex of the root. The collenchyma and fiber cells play a supportive and protective role. The parenchyma cells of the cortex serve as storage areas, although in the stem the parenchyma cells in the pith have the primary storage role.

C. Epidermis

The outer walls of the cells of stem epidermis are coated with a waxy substance, *cutin*, which serves to waterproof them. In the stem, then, the epidermis is concerned with the prevention of water loss, as opposed to its function of water absorption in the root.

PERTINENT FACTS

- There are two basic stem types, woody and herbaceous.
- The stem and root have many structures in common.
- Cambium is a specialized type of meristematic tissue.
- The age of a tree can be determined by counting its annual rings.

5. Physiology of the stem

Again stressing continuity and interrelation of functions, the teacher should build upon previously developed concepts. Students should have little difficulty in understanding most of the functions of the stem from their own textbooks and library readings. Special emphasis should be given to conduction (*translocation*) in the stem. A review of root pressure (Lesson 11) is a good starting point.

Students should be reminded that xylem is the tissue through which upward movement of materials takes place, while phloem conducts materials downward. *What force causes water to move upward in a plant?* As root pressure was previously discussed, your students will probably suggest this as the answer. The teacher might then pose the question, *Does the theory of root pressure adequately explain the movement of water to the top of a tall tree?* The inadequacies of root pressure as the sole force involved can then be examined. Here, the teacher should bring in the *cohesion* theory, sometimes also known as the *shoot tension* theory. One factor to be considered here is the high cohesive force between water molecules, which allows water to be "pulled" up a narrow tube. A second factor would be the *source* of this pulling force. It is now believed by plant physiologists that this pull occurs through the processes of diffusion and osmosis. At this point, have student volunteers briefly review the mechanisms of each. As water is lost through evaporation and cell

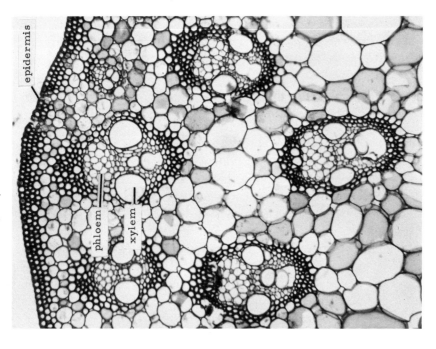
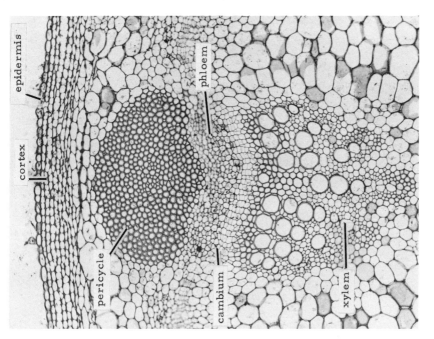

Figure 12-2. Internal Stem Anatomy (Dicot and Monocot).
Courtesy Carolina Biological Supply Company

usage, the contents of the cells become *hypertonic* (more concentrated) to the surroundings, and water is pulled into them. The total effect of this pulling at each cell provides the needed force.

Tracing the pattern, from water loss at the leaves, along the xylem, it can be seen that water must ultimately be brought in from a source outside the plant. The force of shoot tension will pull water into the roots from the surrounding soil. This process, known as passive absorption, will occur when active water absorption by the roots is insufficient. Thus a continuous column of water is maintained in a living plant.

The translocation of food from the leaves downward, occurs through the phloem. No satisfactory explanation of the mechanism involved has yet been developed. If you wish to go into greater depth, the theories of *interfacial flow, cytoplasmic streaming,* and *mass* flow might be explored.

A simple demonstration can be performed during lecture to illustrate water conducting tubes. A freshly cut petiole of a celery leaf should be placed in a beaker of water to which a small quantity of red ink has been added. After a short period of time (15 minutes), the color of the ink will be visible throughout the stalk. If the petiole is then cut near the top and squeezed gently, some of the red fluid will flow out. This demonstration indicates the location of the xylem tubes. In order to observe the tubes in cross-section, the petiole should be sectioned diagonally with a sharp single edge razor blade. The small, dark red spots are the actual tubes. (It should be pointed out that the celery petiole is a fleshy one and acts like a stem.)

6. Economic importance

Some uses of stems which students might volunteer include wood, food (potato, sugar cane), turpentine, and cork. Others are fiber (flax), dyes, perfume (sandalwood), quinine, chicle (chewing gum), rubber, and spice (cinnamon).

PERTINENT FACTS

- Stems produce and support flowers and leaves, act as storage areas, and conduct materials.
- Root pressure and shoot tension are partly responsible for the translocation of fluids through stems.

POSSIBLE QUIZ

1. Discuss the arrangement of tissues in the stem and compare them with those in the root.
2. If you were asked to identify a stem, what characteristics would you use to determine whether it were woody or herbaceous?

3. What is the difference between a node and an internode? Between a lenticel and a leaf scar?
4. In what ways are the steles of the root and the stem different? What are the functions of parenchyma and collenchyma in the cortex?
5. How can you account for the movement of fluids to the top of tall plants?

READINGS

Biddulph, O., and S. Biddulph, "The Circulatory System of Plants," *Scientific American*, February, 1959.

Bold, H.C., *Morphology of Plants*. New York: Harper and Row, Inc., 1967.

Greulach, V., "The Rise of Water in Plants," *Scientific American*, October, 1952.

Hales, S., "The Pressure of Sap in Plants," in *Moments of Discovery*, Vol. 11, G. Schwartz and P. Bishop (eds.). New York: Basic Books, Inc., 1958.

Zimmermann, M.H., "How Sap Moves in Trees," *Scientific American*, March, 1963.

FILMS

"Multicellular Plants, Part III: Stems." 27 minutes, sound, color, $8.15. McGraw-Hill Book Co., Text-Film Division, 330 W. 42nd Street, New York, N.Y. 10036.

MODELS

1. Prepared models of a dicot stem, bio-plastic mounts, microexplano mounts, and microscope slides of stems may be obtained from Ward's Natural Science Establishment, Inc., P.O. Box 1712, Rochester, New York 14603.
2. Preserved specimens of stems may be obtained from General Biological Supply House, Inc., 8200 South Hoyne Avenue, Chicago, Illinois 60620.

Lesson 13

THE LEAF AND FOOD MAKING

Lesson time: 45 minutes
Laboratory time: 90 minutes

AIM

To show the interrelationships of leaf structure and function, with emphasis on photosynthesis.

MATERIALS

Microscope; Glass slides; Cover slips; Leaves of spinach, coleus, geranium, elodea; Geranium and coleus plants; Iodine-water solution; Acetone; Petroleum ether; Alcohol; Electric hot plate; Forceps; Flat dish; Filter paper; Foil, carbon paper, or blotting paper; Paper clips; Glass rod; Adhesive tape; Beaker of water; Celery petiole.

PLANNED LESSON

1. External structure of a leaf

Students should be asked to bring various types of leaves to class; or you can supply them with leaves from potted plants. While observing these, students should be made aware of the various gross structures. The *blade* of the leaf is the entire flat expanded portion which is attached to the leaf stalk or *petiole*. Here the teacher might show his students a petiole of celery with the leaves attached. It can be pointed out that the "stalk" is actually a leaf petiole. A system of *veins*, in a characteristic pattern, can be seen on the surface of the blade. These extensions of xylem and phloem (vascular bundles) branch from the *midvein*, which is continuous with the petiole. The leaf traces of the petiole are the actual extensions of xylem and phloem which connect petiole and stem. *What are branch traces?*

2. Internal structure of a leaf

Reinforcing what has previously been learned about vascular tissues, the teacher can again stress that xylem and phloem are found in certain areas which, in close proximity, are separate and distinct. Specifically, leaf xylem composes the upper portion of veins, while the lower portion is composed of phloem. Parenchyma cells, sometimes with collenchyma and fiber cells, surround these veins. *What functions do these tissues perform?*

The major portion of the leaf consists of chloroplast-containing parenchyma, sometimes called *chlorenchyma*. This is the chief food making tissue of the plant and is known collectively as the *mesophyll*. The teacher should sketch internal leaf structure on the chalkboard and point out these layers. Figure 13-1, illustrating the cross-section of a leaf, will be useful in this respect. It should be pointed out that the *palisade tissue* of

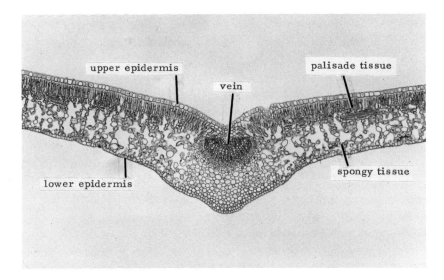

Figure 13-1. Leaf Cross Section.
Courtesy Carolina Biological Supply Company

the mesophyll is under the upper epidermis and consists of two or three cell layers. The cells of this layer are long and are at right angles to the surface of the leaf. The *spongy tissue* is between the palisade cells and lower epidermis, and is composed of irregularly shaped and loosely packed cells. This loose arrangement of cells creates many air spaces.

The epidermis surrounds the entire leaf. Typically, the epidermis lacks chlorophyll and contains cutin. *What purpose might be served by the transparent nature of the epidermis?* Scattered among the epidermal cells are specialized *guard cells* which regulate the size of the pores or *stomata*. *What is the advantage of having a thicker cuticle and fewer stomata on the upper surface of a leaf?*

3. Photosynthesis

Stated simply, photosynthesis is the food making process of plants in which sugar and oxygen are produced from water, carbon dioxide, and energy in the form of light, in the presence of chlorophyll. The teacher should stress the importance of photosynthesis to animals as well as to plants. *What consequences might be expected if all plant life were destroyed?*

Reference should be made to the role of chlorophyll in photosynthesis. Briefly, students should be familiarized with the functioning of a *catalyst*. Chlorophyll may be used as a specific example. It is unnecessary to go into great detail here. The points which should be mentioned include that a biological catalyst, or *enzyme*, is a chemical which speeds up, or makes possible, some chemical change; that a very small quantity of enzyme is needed for the change; that the enzyme is not used up during the change and remains chemically the same afterward, so that it can be reused a number of times before "wearing out"; and that each enzyme has a single specific function within a chemical reaction (the Lock and Key Theory). *Why is "lock and key" an appropriate name for this theory?*

Students should be aware that, in addition to the green chlorophyll, many plants also contain *carotene*, a deep orange pigment, and *xanthophyll*, a bright yellow pigment. These, together with other accessory plant pigments, play a role in leaf coloration. Little is known concerning the specific functions of these pigments. *Why do leaves change color as they die in the fall?*

The teacher should bring out the fact that there are two main phases in the process of photosynthesis, the *light reaction* and the *dark reaction*. The light reaction (Hill Reaction) is a light-driven reaction which consists of the photochemical decomposition of water. This decomposition

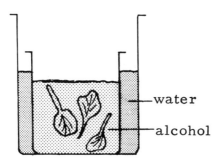

Figure 13-2. Water Bath Double Boiler.

of water by light is known as *photolysis*. Students should always be encouraged to understand the terms used. "Photochemical" is self-explanatory, and students should be able to indicate the materials suggested. If the teacher explains the meaning of word parts, such as *lysis* or breaking-apart, they will develop an ability to analyze terminology, and apply what they have learned to other cases. *What do the following terms imply: Lysol, hemolysis, plasmolysis, cytolysis?*

The dark reaction is so named, not because it *must* take place in the dark, but rather because it *can* occur in the dark. That is, light is not essential. The dark reaction involves the fixation of carbon dioxide to produce organic compounds such as sugar. The hydrogen produced by photolysis combines with the carbon and oxygen of carbon dioxide. The relationship of the two stages should now be evident. The general formula for the overall process of photosynthesis is:

$$\underset{\text{carbon dioxide}}{6CO_2} + \underset{\text{water}}{6H_2O} + \text{enzymes} + \underset{\text{light energy}}{E} \longrightarrow \underset{\text{glucose}}{C_6H_{12}O_6} + \underset{\text{oxygen}}{6O_2}$$

4. Laboratory activities

A. Starch Making by Green Leaves

A simple test for the presence of starch in leaves can be performed by your students. Since the test will use an iodine solution to "decolorize" the leaves by removing chlorophyll, geranium or coleus leaves can be used. The leaves should be boiled in water for 5 minutes to break down the cell walls. Next, an alcohol bath is needed to remove the alcohol-soluble chlorophyll. It is suggested that alcohol baths be set up in one restricted area only, where students can be closely supervised. *Always use a beaker half full of alcohol which has been placed within a larger beaker of water* (double boiler), as illustrated in Figure 13-2. If possible, an electric hot plate should be used to avoid open flames. Caution your students before beginning that alcohol fumes ignite readily. In about 5 minutes, the leaves should be decolorized.

The leaves should be carefully removed from the alcohol with forceps, and placed in a flat dish. An iodine-water solution (Lugol's may be used if avaliable) should be poured over the leaves and allowed to remain for 5–10 minutes. After rinsing, students should observe the characteristic black areas which indicate the presence of starch.

B. The Effect of Light on Carbohydrate Synthesis

As a modification of the preceding activity, students might use geranium plants which have been kept in a dark place for at least 12 hours immediately prior to use. Part of several leaves should be covered with pieces of opaque material (blotting paper, carbon paper, foil). They might use paper clips, pins, or tape to fasten the covering material firmly. Plants should be placed in direct light and allowed to remain there for at least 24 hours. The leaves may then be tested for the presence of starch.

In order to perform these tests during the same laboratory session, the preliminary preparation of the leaves can be performed a day or two earlier, during a class period. Students should recall results and suggest possible explanations. *What might be used as a control so that this activity could be a true experiment?* Students should be encouraged to create and use controls whenever possible.

C. Chromatography

Students can set up a simple chromatography analysis. The apparatus should be assembled as illustrated in Figure 13-3.

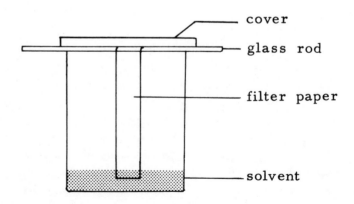

Figure 13-3. Chromatography Apparatus.

The strip of filter paper should be hung from a glass rod and covered to prevent evaporation of the solvent.

A small quantity of spinach should be ground in a mortar or electric blender with acetone (5gm of spinach to 10cc of acetone). A spot of this mixture should be placed on the filter paper strip so that it is approximately one inch above the level of the solvent. The solvent should include equal parts of acetone and petroleum ether.

Better results will be obtained if the spot is allowed to dry and a second drop added on top of the first. In either case, students will observe the movement of solvent up along the filter paper, and the resultant movement of color upward from the spot. As time progresses, they will see that the spot separates into different colors. This is an indication of the presence of various plant pigments, and is based upon the difference in the diffusion rates of their molecules. Students can identify these pigments by observing their diffusion. The fastest rate of diffusion occurs in carotene, followed by xanthophyll 1 and 2, and chlorophyll a and b, in descending order. Interested students will find much library information on the varied uses of chromatography.

D. Observation of Chloroplasts

Students should prepare wet-mounts of fresh Elodea leaves and observe the movement of their chloroplasts. They will also be able to observe cytoplasmic streaming. They should be reminded that they are expected to record their observations of all laboratory work.

PERTINENT FACTS

- The chief food manufacturing tissue of the leaf is mesophyll.
- Leaf epidermis must be transparent so that light can reach photosynthetic cells within the leaf.
- During the light reaction of photosynthesis, the hydrogen and oxygen of water are separated in the presence of light and chlorophyll.
- In the dark reaction of photosynthesis, the hydrogen released during the light reaction is "fixed" with carbon dioxide to produce carbohydrates.
- Chlorophyll is one of many enzymes essential to photosynthesis.
- The various pigments found in chloroplasts include carotenes and xanthophylls as well as chlorophyll, and may be separated by chromatography.

POSSIBLE QUIZ

1. What are the functions of leaf veins? How does the structure of the leaf relate to the stem and root?
2. What is the function of leaf epidermis? Contrast this with root and stem epidermis.
3. What are some characteristics of an enzyme? How do these characteristics allow for greater efficiency? What enzyme is important in photosynthesis?
4. Briefly explain photosynthesis, including the two main stages.
5. Describe a simple activity demonstrating the role played by light in photosynthesis. What principle does chromatography involve? Suggest some possible applications of chromatography.

READINGS

Arnon, D.I., "The Role of Light in Photosynthesis," *Scientific American*, November, 1960.

———, "Photosynthesis: Facts and Concepts," *Science Teacher*, December, 1967.

Galston, A.W., *The Life of the Green Plant*. Englewood Cliffs, New Jersey: Prentice-Hall, Inc., 1964.

Geisert, P., "Experiment in Photosynthesis: The Hill Reaction," *Science Teacher*, December, 1967.

McElroy, W.D., and B. Glass (eds.), *Light and Life*. Baltimore: The Johns Hopkins Press, 1961.

Rabinowitch, E.I., "Photosynthesis," *Scientific American*, August, 1948.

———, "Progress in Photosynthesis," *Scientific American*, November, 1953.

Wald, G., "Light and Life," *Scientific American*, October, 1959.

FILMS

"Biological Techniques: Paper Chromatography." 14 minutes, sound, color, $5.15. AIBS with Thorne Films, Inc., 1220 University Avenue, Boulder, Colorado.

"Multicellular Plants, Part III: Chlorophyll." 26 minutes, sound, color, $8.15. McGraw-Hill Book Co., Text-Film Division, 330 W. 42nd Street, New York, N.Y. 10036.

"Multicellular Plants, Part III: Leaves." 27 minutes, sound, color, $8.15. McGraw-Hill Book Co, Text-Film Division.

"Multicellular Plants, Part III: Role of the Green Plant." 26 minutes, sound, color, $8.15. McGraw-Hill Book Co., Inc., Text-Film Division.

"Plant Life: Photosynthesis." 23 minutes, sound, color, $7.65. Encyclopedia Britannica Films, Inc., 1150 Wilmette Avenue, Wilmette, Illinois 60091.

"The Riddle of Photosynthesis." $12\frac{1}{2}$ minutes, sound, bw. Handel Film Corporation, 6926 Melrose Avenue, Hollywood, California 90038.

MODELS

1. Students can make impressions of the veination pattern of various types of leaves using such materials as wax, clay, or foil.
2. Charts, display mounts, microscope slides, and models of leaves may be obtained from General Biological Supply House, Inc., 8200 South Hoyne Avenue, Chicago, Illinois 60620.

Lesson 14

THE LEAF AND RESPIRATION

Lesson time: 45 minutes
Laboratory time: 90 minutes

AIM

To develop the principles involved in the exchange of gases between the plant and its environment.

MATERIALS

Microscope; Glass slides; Cover slips; Plants—Rheo discolor, geranium, and cactus; Test tubes and rack; Graduated cylinder; Single edge razor blade; Petroleum jelly; Plastic bags.

PLANNED LESSON

1. Stomata

Having been introduced to the location and structure of stomata, your class should now consider their function. It has previously been indicated that the stomata are actually pores in the leaf epidermis. Using information developed in Lesson 13, students should be able to explain the relationship between guard cells and stomata. *How is the size of a pore controlled?* Here again the teacher should encourage his students to under-

The leaf and respiration

stand terminology. "Stome," meaning mouth, will be seen in many terms, as in hypo*stome*, cyclo*stome*, and *stom*ium.

A brief look at the role that these pores play in gas exchange might be brought in here. *How might gases such as oxygen, carbon dioxide, and water vapor enter and leave the leaf?*

2. Respiration

A point which must be stressed by the teacher and clearly understood by the students is that respiration in plant cells is identical with respiration in animal cells. That is, during the energy cycle which occurs in the mitochondria, oxygen is used and carbon dioxide is given off. A great deal of confusion results when students learn that plants give off oxygen and take in carbon dioxide during the day, and give off carbon dioxide and take in oxygen at night. This is the result of a higher rate of photosynthesis during the day than at night. It should be pointed out that respiration occurs 24 hours a day in plant cells.

During the day, photosynthesis occurs at a high rate, and therefore requires a great deal of carbon dioxide. The carbon dioxide which is an end product of respiration may be picked up and used during photosynthesis. This source, however, is inadequate and additional carbon dioxide must diffuse in through the stomata. *What will the net result be?* Conversely, photosynthesis produces oxygen. Some of this oxygen will be used as a source for the process of respiration, while the excess oxygen diffuses out of the leaves into the environment.

Exactly the same logic can be used to explain what happens at night, when the rate of photosynthesis is greatly reduced. This means that far less carbon dioxide is needed. In terms of chemical activity, the process of respiration is now greater than that of photosynthesis. Applying what they have just learned, students should be able to explain that respiration requires oxygen, which now is not available as an end product of photosynthesis, so that all needed oxygen must be obtained by diffusion from the environment. The carbon dioxide which is an end product of respiration is not used in large quantities for photosynthesis, so it diffuses out into the environment. Thus, the exchange of gases between the leaf and the environment is the net result of two separate processes occurring within the leaf.

3. Transpiration

The evaporation of water from a plant is called transpiration. This material should serve as the final link in the chain of water movement from root to leaf. A brief review of the concepts concerning water movement in roots and stems (Lessons 11 and 12) should be presented as a unifying theme. Transpiration has, in fact, previously been introduced during the discussion of shoot tension. The amounts of water lost through transpiration are significant. An acre of corn transpires 325,000 gallons of water during a single growing season! *What effect might transpiration have on climate? What environmental factors influence transpiration?* Students should be able to suggest that temperature would be a factor, as would humidity and light. *How is wilting related to transpiration?*

4. Laboratory activities

A. STOMATA MAY BE OBSERVED

Stomata may be observed by having students prepare a wet mount of the epidermis of the lower surface of a leaf. A suggested plant would be Rheo discolor, a succulent whose lower epidermis can be peeled off easily.

B. STUDENTS CAN PLAN AND CONDUCT AN EXPERIMENT

Students can plan and conduct an experiment to answer each of the following questions:

1. From which leaf surface does a plant transpire most?
2. What is the effect of temperature on transpiration?

Students should work in groups of 3 or 4 and should consider the following:

1. variables;
2. assumptions made;
3. collection, recording, and reporting of data.

Each group of students should be supplied with the following materials: several potted plants, such as geraniums, from which cuttings can be made; test tubes and rack; petroleum jelly (a thin layer of petroleum jelly spread on a leaf will reduce water loss); a single edge razor blade; and a graduated cylinder.

After the experiment has been performed, reports should be presented in proper written form.

C. AS A DEMONSTRATION

As a demonstration, a potted plant should be adequately watered, and then covered with a plastic bag securely tied *under* the flower pot. As gases and water are now contained within a sealed environment, the plant should flourish indefinitely. *What factors might effect the functioning of this system?* The teacher might point out that this is an excellent way to protect houseplants if they must be left untended for long periods.

D. AN EXPERIMENT CAN BE DEVELOPED

An experiment can be developed from the preceeding demonstration. Using a houseplant and a cactus (all other factors should be as equal as possible—such as pot size, plant size, and amount of water provided), water both plants and tie plastic bags over each, securing each at the base of the stem just above the soil. Students should observe them for evidence of transpiration.

PERTINENT FACTS

- Gases enter and leave a plant through stomata.
- Respiration is a continuous process in all living cells.
- During respiration oxygen is used and carbon dioxide is given off.
- During photosynthesis carbon dioxide is used and oxygen is given off.
- The exchange of oxygen and carbon dioxide between the leaf and the environment is the net result of the gas requirements of photosynthesis and respiration.
- Transpiration is a loss of water vapor from the plant.

POSSIBLE QUIZ

1. What functions do stomata have in respiration and photosynthesis? Are the functions similar in each?
2. How does gas exchange in respiration differ from gas exchange in photosynthesis?
3. Explain why plants take in carbon dioxide during the day and give it off at night.
4. What is the interrelationship between photosynthesis and respiration?
5. Explain how temperature, light, and humidity affect transpiration. At which leaf surface does most of the transpiration occur?

READINGS

Bonner, J., and A.W. Galston, *Principles of Plant Physiology*. San Francisco: W.H. Freeman and Co., 1952.

Heath, O.V.S., "The Water Relations of Stomatal Cells and the Mechanisms of Stomatal Movement," *Plant Physiology*, Vol. II. New York: Academic Press, 1959.

Siekevitz, P., "Powerhouse of the Cell," *Scientific American*, July, 1957.

Zimmermann, M.H., "How Sap Moves in Trees," *Scientific American*, March, 1963.

FILMS

"Multicellular Plants, Part III: Role of the Green Plant." 26 minutes, sound, color, $8.15. McGraw-Hill Book Co., Text-Film Division, 330 W. 42nd Street, New York, N.Y. 10036.

"Multicellular Plants, Part III: Stems." 27 minutes, sound, color, $8.15. McGraw-Hill Book Co., Text-Film Division.

Lesson 15

FLOWERS, FRUITS, AND SEEDS

Lesson time: 45 minutes
Laboratory time: 45 minutes

AIM

To develop concepts related to the reproductive cycle of flowering plants.

MATERIALS

Microscopes; Glass slides; Fresh flowers or model of a flower; Pollen grains; Castor and lima beans; Fresh fruits—such as apples, squash, cherries, olives, cucumbers, or tomatoes; Petri dishes; Hand lens; Sharp knife or scalpel; Blotting paper; Small brushes; Mayer's albumin fixative: glycerin, phenol crystals, cheesecloth, egg albumin.

PLANNED LESSON

Although flowering plants are capable of *asexual* reproduction, as in the production of new plants by bulbs, runners, tubers, and rhizomes, this lesson concentrates on the *sexual* reproduction of plants.

1. Flowers

In angiosperms, flowers are organized groups of modified organs which are the characteristic reproductive structures of the plant. The teacher should make use of a commercially prepared model of the parts of the flower or, if he prefers, students can each bring in a fresh flower to examine. Students should examine their specimens and identify the different flower structures as the teacher points them out on the model. Beware of composite flowers; make sure students bring in "simple" flowers.

It will be seen that the various flower parts are attached to a stem-like structure known as the *receptacle* or torus. In general, the parts of the flower are divided into four concentric whorls. The whorl closest to the receptacle is known as the *calyx*, and is composed of individual parts known as *sepals*. *What do sepals resemble?* The petals of the flower constitute the second whorl. Collectively, this group is known as the *corolla*. Mention should be made to your students that plants which are primarily pollinated by insects have conspicuous and brightly colored petals. *Why?* The third group is composed of the *stamens* of the plant. Each stamen consists of a slender filament supporting an *anther*. Stamens are sometimes known as *microsporophylls*, because they produce the microscopic spores which are the pollen. In effect, the anther is a group of pollen capsules within which pollen grains are produced. The fourth layer comprises the *carpels*. In the majority of flowers, the carpels are joined in a single structure known as the *pistil*. The pistil is a compound organ composed of *stigma*, *style*, and *ovule chamber* (ovary). The pistil is sometimes known as the *macrosporophyll*. *Why is this term used?* Students should use hand lenses to examine these structures, and should be asked to identify the "male" and "female" reproductive structures. *Which produces the ovules? Which produces the fertilizing cells?* Figure 15-1 may be placed on the chalkboard to illustrate the various flower parts, including the reproductive organs.

Some species of flowers do not have both male and female structures within a single flower. If only stamens are present, the flower is said to be *staminate* or "male." *Pistillate* flowers, also called "female," contain only pistils. In some species, pistillate and staminate flowers are found on a single plant, while in others the entire plant will have only one or the other. The ginkgo, or maidenhair tree, is an example of a species in which male trees and female trees are produced. Other examples would include holly, willow, pear, and cottonwood.

2. Fertilization

The teacher might here review the concepts of zygote formation. Students should have little difficulty in outlining a logical sequence of events

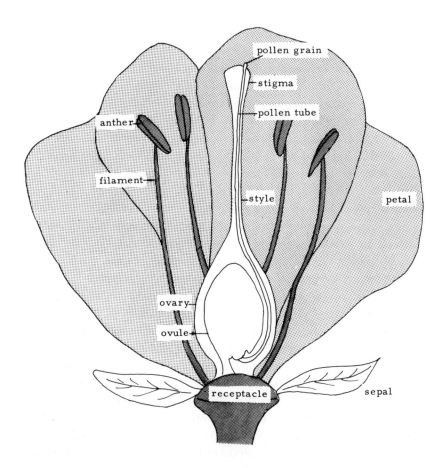

Figure 15–1. Flower Parts.

which might occur during fertilization. With the teacher's guidance, an accurate outline can be developed:

A. Pollination

Insects, wind, water, and birds, as agents of pollination, should be mentioned. Self-pollination should also be included. The teacher should encourage his students to help in the pollination of classroom plants when flowers develop. Small brushes or other appropriate devices can be used to transfer the pollen to the stigma.

B. Pollen Tube Formation

After a pollen grain lands on the sugary secretion at the top of the stigma, it begins to grow downward through the tissues of the style until it reaches the *ovulary*.

A simple laboratory activity which can be included here would involve the growth and observation of pollen tubes. Have students place a thin layer of Mayer's albumin fixative (25 cc of glycerin, .5 g phenol crystals—carbolic acid, 25 cc egg albumin, filter through cheesecloth) on the surface of a clean glass slide. A few grains of pollen should be placed on the slide, and the slide then placed in a warm, humid location. Humidity may be insured by placing the slide in a covered petri dish in which a ball of moist cotton or moist blotting paper has also been placed. Pollen tubes can be observed after several hours. Students should use microscopes for thorough examinations. (Volunteers can be selected a day earlier, so that they can prepare the materials to be used in class time. Allow 30 minutes for preparation.)

C. Zygote Formation

Fertilization occurs when the pollen tube completes its downward growth through the style to the ovulary. *What process (cell division) leads to the production of pollen? Ovules?*

An interesting film which could serve as a review of the lesson thus far would be "Flower Functions."

3. Fruits

To graphically illustrate the transformation of the ovule chamber into a fruit, the teacher may make use of a number of fruits which can be dissected in class. Apples, cucumbers, squash, tomatoes, olives, and cherries can be used. It should be apparent that the edible portion of these fruits is derived principally from the ripened ovary. The teacher should ask his students to consider whether the term "fruit," as commonly used, is in fact scientifically accurate.

4. Seeds

Germination or sprouting has previously been observed in a series of activities (Lesson 11). These activities may be repeated here so that seed growth can be observed. Castor and lima beans may be used and are especially valuable, since their large size permits easy observation.

In the laboratory, students should examine similar beans which have been soaked in water overnight. They should use a sharp instrument to carefully split the seed into two parts. The teacher should demonstrate this before the students begin. It will be noted that the seed is composed of two halves or *cotyledons*. Using a hand lens, students can identify the embryonic leaves or *plumules*. The part between the plumules will be the *epicotyl* or main shoot. At the end away from the epicotyl is the *hypocotyl*, where the embryonic root is formed. Germinated seeds should be observed at this time to obtain a clearer picture of seed structures. It will further be noted that, as the seeds germinate, the cotyledons shrink, eventually dry up, and then fall off. *What is the function of cotyledons?* Figures 15-2 and 15-3 may be placed on the chalkboard during lecture-laboratory for reference. These figures illustrate the parts of germinating corn and bean seeds. The film "Seeds and Germination" may also be used.

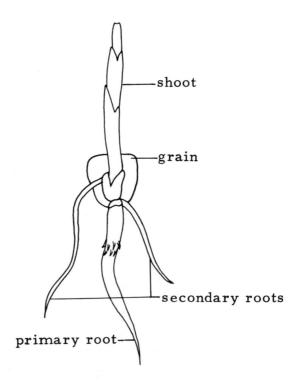

Figure 15–2. Corn Germination.

PERTINENT FACTS

- Flowers are reproductive plant structures.
- Staminate flowers are male flowers and pistillate flowers are female flowers.
- Fertilization occurs when the pollen tube grows downward through the style, allowing for the union of sperm and egg.
- The ovary develops into the fruit, whereas the ovules become the seeds.
- Seeds are embryo plants.

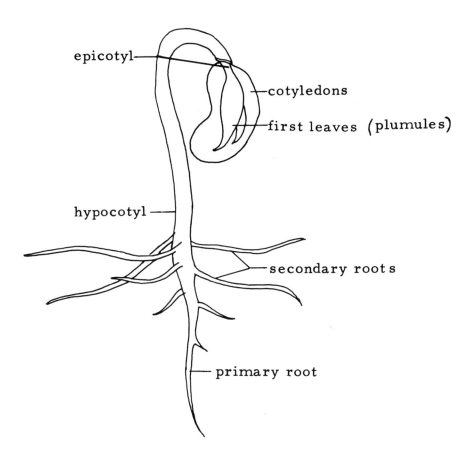

Figure 15-3. Bean Germination.

- Germination is the sprouting of seeds under favorable environmental conditions.

POSSIBLE QUIZ

1. Sketch a flower and identify its major parts.
2. What flower parts would be expected to have the haploid number of chromosomes? Why?
3. Describe how fertilization occurs in flowers. From what plant structures do the fruit and seed develop?
4. There are two kinds of fruit, dry and fleshy. From the following list, select those which are fruits, and indicate whether they might be dry or fleshy. Cucumber, lemon, squash, pumpkin, onion, banana, acorn, lima bean, olive, chestnut, corn, pea, rice, sweet potato, pineapple, strawberry, lettuce, broccoli, brussels sprouts, string beans, watermelon.

READINGS

Biale, J.B., "The Ripening of Fruit," *Scientific American*, May, 1954.

Echlin, P., "Pollen," *Scientific American*, April, 1968.

Esau, K., *Anatomy of Seed Plants*. New York: John Wiley and Sons, 1960.

Grant, V., "The Fertilization of Flowers," *Scientific American*, June, 1951.

Koller, D., "Germination," *Scientific American*, April, 1959.

FILMS

"Flowers at Work." 11 minutes, sound, bw, $2.15. Encyclopedia Britannica Films, Inc. 1150 Wilmette Avenue, Wilmette, Illinois 60091.

"Multicellular Plants, Part III: Flower Functions." 27 minutes, sound, color, $8.15. McGraw-Hill Book Co., Text-Film Division, 330 W. 42nd Street, New York, N.Y. 10036.

"Multicellular Plants, part III: Flower Structure." 30 minutes, sound, color, $8.15. McGraw-Hill Book Co., Text-Fim Division.

"Multicellular Plants, Part III: Seeds and Germination." 24 minutes, sound, color, $8.15. McGraw-Hill Book Co., Text-Film Division.

MODELS

1. Charts of flowers and seeds, models of a flower, microscope slides and plastomounts of flowers, fruits, and seeds may be obtained from Carolina Biological Supply Co., Burlington, North Carolina 27215.
2. Models of fruits and seeds may be obtained from Ward's Natural Science Establishment, Inc., P.O. Box 1712, Rochester, New York 14603.

Lesson 16

PLANT RESPONSES

Lesson time: 180 minutes
Laboratory time: 90 minutes

AIM

To develop an integrated study of the many phases of plant reactions, including the mechanisms involved in growth and turgor movements.

MATERIALS

Oat, bean, and corn seedlings; Begonia, morning glory, azalea, and geranium plants; Petri dish; Test tubes; One-hole rubber stoppers; Paper cups; Distilled water; Blotting paper; Aluminum foil; Cheesecloth; Single edge razor blades; Cotton or sand; Mineral solutions; Indolacetic acid: indolacetic acid, ethyl alcohol, lanolin paste.

PLANNED LESSON

Among higher plants, two general types of movement might be pointed out by the teacher: *growth movement* and *turgor movement*.

1. Growth movements

Growth movements are characterized as being slow and nonreversible. Student discussion centering around their knowledge of plant growth should lead to some understanding of the rate of growth in plants. *Name some plants which grow relatively fast, and others which grow slowly.*

Scientists studying the nature of plant growth have found that the growth of a plant or plant part is due to the presence of a diffusable substance or hormone found in plant tissue. These substances are also known as *auxins*. Due to the presence of hormones, an unequal lengthwise growth of cells occurs in one part of a plant as opposed to another. These natural plant hormones are produced primarily by *meristem tissue* in the growing parts of the plant. An example of the unequal growth of cells resulting in plant "movement" would be the bending of a plant stem in the direction of light.

Students should be encouraged to develop a hypothesis to explain this bending. They will generally suggest that the sunlight causes the cells exposed to it to grow more rapidly, thus causing the stem to bend. The teacher should point out that, since growth is considered to be a longitudinal lengthening of cells, the edge of the stem *away* from the light would have to have longer cells, since the curvature is greater on that edge. You will find Figure 16-1 most helpful in clarifying this point.

Once students are aware of the growth pattern, they should attempt to develop an explanation of the relationship of sunlight to auxins. They should understand that smaller concentrations of auxin are produced within those cells receiving a greater amount of light. As auxin production is inhibited, the rate of cell growth decreases. Since the quantity of auxin produced within the cells on the darker side of the stem is relatively greater than the quantity produced by inhibited cells, an unequal growth rate can be observed. This unequal growth pattern causes the bending of the stem. This bending in response to light is called *phototropism*.

2. Effects of auxins

Although the mechanisms of the functioning of auxins are not yet fully understood, some of the cell processes affected have been identified. These include cell wall formation, enzyme action and respiration, protein metabolism, and the absorption of materials.

Auxins are responsible for the falling of leaves in autumn. The *abscission* of leaves is the result of the reduction in the quantity of auxins reaching the base of the petiole. This causes a change in the ratio of auxin in the cells on the two sides of the petiole base. This auxin imbalance triggers a differentiation of cells which develop into the abscission layer. This is the weakened area at which the petiole snaps, causing the leaf to fall.

What explanation can you propose for the falling of ripe fruit from the stem? What causes fruit to drop before it is ripe? It should be mentioned that such premature fruit drop may be prevented by spraying with solutions of auxins. The development of the fruit itself is dependent upon an adequate

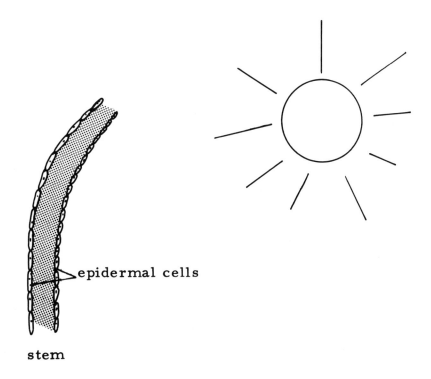

Figure 16-1. Stem Growth Toward Sun.

supply of auxin. Generally, this supply of auxin is produced by the developing embryo within the seed. As a result, fruits fail to develop unless flowers are pollinated. *How might seedless fruits develop?* Students should be led to use their knowledge concerning auxins to answer this. Since they know that premature dropping can be prevented by spraying, they should be able to extend this concept.

Other uses for auxins include greenhouse production of fruits and vegetables, regardless of season. One of the most valuable plant hormones used is *gibberellic acid.* One of the more striking effects of gibberellin is the promotion of flowering. Using previous knowledge of auxin functioning, students should be able to trace the entire pattern of greenhouse fruit production. Gibberellin produces unusually large plants. Cabbages treated with gibberellic acid bloom much earlier than untreated ones and grow much taller.

3. External factors affecting growth

From previous investigation, students can indicate that light would be an external factor effecting growth. Other factors they can list would include water, temperature, minerals, oxygen, and carbon dioxide.

A. Light

Light and its effects on auxin production is one factor previously discussed. Other phenomena involving light may also be observed in plants. Students should be encouraged to observe and care for classroom plants throughout the year. The teacher should not assume that they will have had such experiences outside the classroom. By observation they should be able to note that some types of plants do well in indirect light, while others require full sunlight. They might observe the effects of varying light concentration on individual plants by changing their locations. Here again, students should attempt to understand the meaning of terms. *Photophobic* plants grow best in indirect light and will do poorly in bright sunlight. (*What is a "phobia?"*) Photophilic plants (-philic from the Greek word for "friend") require a great deal of light for good growth.

The length of daily exposure to light, or the *photoperiod* (*term derivation?*), greatly effects growth, development, and reproduction. The most dramatic effect is on flowering. Each plant species requires a specific photoperiod for the initiation of flowering. The teacher might use the Poinsettia as a familiar and interesting example. Known as a "short-day" plant, the Poinsettia will not flower if its daily photoperiod is longer than 13–14 hours. In the North Temperate Zone, this coincides with the length of days during winter, and so Poinsettias are well-known Christmas flowers. Many people who receive Poinsettia plants as gifts are disappointed because they do not flower the following year. It has been found that the light from night-time illumination, or from a television set in an otherwise darkened room, can lengthen the photoperiod and thus inhibit flower production. Other short-day plants are salvia, dahlias, asters, and chrysanthemum.

"Long-day" plants include delphenium, gladiolas, clover, and corn. These require more than 14 hours of light daily. A third category of plants exists; these are the "day-neutral" or *indeterminate* plants. These flower under long or short day conditions. They include tomatoes, sunflowers, carnations, and dandelions. *How does photoperiodism affect the world distribution of plants?*

It should be emphasized that animals are also affected by length of day. Examples of the effects include seasonal breeding, change in fur color, and possibly bird migration.

B. Other Factors

The temperature of the soil and surrounding atmosphere influences plants. Although the ideal temperature varies from species to species, temperatures below freezing or near 100°F will generally cause growth stoppage. Most plants survive well within a range of 70° to 90°.

The quantity of water available to a plant will have a marked effect on its growth pattern. Here, too, needs vary from species to species, ranging from the cactus to the water lily.

Minerals are essential to normal growth. The formation of chlorophyll, food synthesis, production of protoplasm, and all other metabolic activities, require various minerals. The lack of minerals will prove detrimental to normal functioning and therefore to growth. Many mineral deficiencies can be identified by a characteristic change in the physical appearance of the plant. These changes include:

1) *Chlorosis.* Lack of chlorophyll. A characteristic yellow or pale green color which is caused by a lack of nitrogen or magnesium.

2) *Necrosis.* Death of parts of the plant. Dead spots on leaves—iron deficiency; edges and tips of leaves die—phosphorous deficiency; buds and roots die—boron deficiency.

3) *Stunted growth.* A deficiency of any plant element will tend to cause retardation of normal growth. Those elements which have the greatest direct effect on growth are nitrogen, phosphorous, boron, and magnesium.

The roles of oxygen and carbon dioxide in respiration and photosynthesis have been explored in depth in Lesson 14. A brief review by student volunteers should be included here.

PERTINENT FACTS

- An auxin is a plant growth hormone.
- These hormones are diffusable substances and exert their effect by regulating cell elongation.
- Gibberellin is a widely known growth stimulator.
- The influence of varying periods of light exposure on plants is called photoperiodism.
- Other factors which effect plant growth include light, minerals, water, and environmental gases.

4. Turgor movement

If the teacher explains that "turgor" is derived from the word which means "to swell," students should be able to develop the meaning of *turgor pressure* and *turgor movement.* Once they understand that the amount of water in the cell creates pressure (turgor pressure), the concepts of turgor movement become easier to understand. Reminding students that water moves into and out of cells by osmosis (as described in Lesson 6), the teacher should ask his students to fully explain the relationship between osmosis and turgor. *How can a piece of wilted celery be restored to crispness?*

Although the structure and function of guard cells has previously been considered (Lesson 14), the opening and closing of the stomata are directly related to turgor pressure within the guard cells. A student volunteer might explain this process in detail.

Whereas growth movements are the result of the effects of specific diffusable substances, turgor movements are the results of osmotic changes occurring in specialized cells. There are three types of cells involved in turgor movement: *receptors, conductors,* and *effectors.* A stimulus is picked up by receptor cells, carried by conductor cells to effector cells, where a change occurs as a result of the stimulus. Turgor movements may be observed in the sensitive plant Mimosa pudica, commonly found in most parts of the United States. Several seconds after the stems of this plant are sharply tapped, the leaves will fold successively, beginning with the leaves nearest the stimulus. This effect can be explained in the following manner.

The stimulus is picked up by the receptor cells of the stem and carried by the vascular tissues of the plant (conductor cells) to the *pulvinus.* This structure is a cushion of cells surrounding the base of the petioles. Under normal conditions, the cells of the pulvinus are turgid and support the leaves in their normal position. When the stimulus reaches these specialized effector cells, the cells become more permeable and lose turgor. This results in a relaxing of the cells, and the leaf folds.

Other examples of turgor movement may be seen in the opening and closing of the leaves of insectivorous plants, such as the Venus flytrap (contact movements). The folding of the leaves of wood sorrel (oxalis) at night in "sleep movements" is another.

PERTINENT FACTS

- Turgor is a term which implies swelling.
- Turgor movements are rapid and reversible whereas growth movements are slow and irreversible.
- Turgor pressure is directly related to osmosis.

5. Tropisms

Tropisms are oriented movements controlled by the position of the stimulus. Most tropisms are growth movements which result from the effects of particular stimuli on auxin production. Movements directed toward the stimulus are known as positive tropisms, while those moving away from the stimulus are known as negative tropisms. Positive tropisms include the previously mentioned phototropism, as it appears in stems and leaves. Negative phototropism is exhibited by roots which grow away from light. Roots exhibit positive *geotropism* in response to the earth's gravity, and positive hydrotropism in response to water. Stems usually exhibit negative geotropism.

Chemotropism is a response to chemicals. An example would be the growth of pollen tubes through the style, and the growth of roots toward water. (Since water is a chemical, this is chemotropism as well as hydrotropism.) *Thigmotropism* is a response to touch, such as the twining of ivy around supporting objects.

PERTINENT FACTS

- Tropisms are movements controlled by the position of the stimulus.
- Most tropisms are growth movements, although some tropisms may involve turgor movements.
- Tropic stimuli exert their effect by influencing auxin production.

6. Laboratory activities

A. Growth

Students should develop a controlled experiment which would support the hypothesis that auxins are produced by the *apical* (tip) meristem.

One possible method would be the preparation of two groups of seedlings, such as oat seedlings. Oat *coleoptiles* (the coleoptile is the first part of the oat plant that emerges from the soil) do well. One group will be the control, while the apical meristem from the other seedlings should be carefully removed with a sharp knife or single edge razor blade. All other conditions must be kept as nearly the same for both groups as possible. Optimum growth conditions should be maintained. The class should observe both groups every 24 hours for the next five days. (If possible this experiment should be begun on a Monday.) *What differences in growth would you expect to find between the two groups? Explain.* Be certain that you record all data accurately and completely.

After observation, students should analyze all data. Points to be brought out are:

1. auxins are produced by the apical meristem;
2. removal of the meristem will, therefore, remove the source of growth hormone;
3. the stem will regenerate meristematic tissue at the tip within 24 hours;
4. the regenerated meristematic tissues produce auxin which is

responsible for the bending of the stems in the experimental group, which can be observed beyond the first 24 hour period.

A second experiment which may be used in the study of growth characteristics is a modification of the preceding one. The removed tips should be exchanged and carefully placed on the remaining stumps. Students should make certain to carefully label each plant, such as A,B,C,D. The labels on stems on which exchanged tips have been placed would indicate the letter of the plant stem and of the tip, such as plant A, tip B. There is no need to attach the tips with foreign substances. Students should collect, record, and carefully analyze all data. Using previous knowledge, they should be able to suggest a possible explanation of the results obtained.

B. Effects of Auxins

Bean seedlings may be used in a similar controlled experiment. Six healthy young seedlings should be divided into three groups. Group A will be the control; Group B will have the apical meristems removed (above the upper set of leaves); Group C will have the apical meristems removed and will be treated with indoleacetic acid in lanolin. The indoleacetic acid preparation should be made as follows: Dissolve 100 mg of indoleacetic acid in 2 cc of ethyl alcohol. Mix this into 10 gm of lanolin, so that it is evenly distributed. This is a 1% indoleacetic acid preparation.

This lanolin paste should be placed on the cut stems of Group C by using a toothpick or other suitable tool so that only a small ball of paste covers the cut tip. Although it is necessary to cover the cut surface completely, excess materials should not come in contact with other parts of the stem. All plants should be maintained under optimum conditions for growth. After careful observations have been made and recorded, data should be analyzed. *What is the function of indoleacetic acid?*

The teacher may want to point out here that indoleacetic acid is actually one of the more common auxins. Student reports on other auxins might be appropriate at this time.

C. External Factors Effecting Growth

1) Photoperiodism

Continuous observation of classroom plants throughout the year can be used to point out photoperiodism. *Why do some plants bloom in winter while others bloom in summer? Is temperature a primary factor?* Interested students can set up specific experiments within the classroom using artificial light sources to devise controlled experiments. These experiments should be based on information presented in this lesson, as well as on outside readings.

The same procedure can be used to test for photophobic and photo-

PLANT RESPONSES

philic plants. Azalea, geranium, begonia, and morning glories are suitable plants to study. Dramatic results will be seen if flowering potted azaleas are used, one in the shade, one in bright sunlight.

2) THE EFFECTS OF MINERALS

The effects of minerals can be demonstrated by growing seedlings in water solutions. The growing of plants in water nutrient solutions is called *hydroponics*. *Why is water used in place of soil?* Seeds should be germinated in a petri dish on moist blotting paper, as described in Lessons 11 and 15. *Why are full-grown plants used?*

After two weeks (germination should be begun in advance), roots will have developed sufficiently and the seedlings can be used. Each seedling will need to be supported over a container of distilled water. A number of suitable techniques work well: a series of test tubes, each containing a single seedling supported by a one-hole rubber stopper; a series of beakers each containing a single seedling supported by aluminum foil stretched across the top of the beaker, in which a suitable hole has been punched (or a piece of cheesecloth similarly prepared); or glass jars, glasses, or paper cups. Students will no doubt offer suggestions of their own.

The control seedlings should be placed in solutions which contain all essential minerals for normal growth. Although many different solutions may be used, only two have been included here.

Knop's Solution (weigh out the following and dissolve in 1 liter of distilled water):

$FePO_4$	trace
KNO_3	0.2 gm
$MgSO_4 \cdot 7H_2O$	0.2 gm
KH_2PO_4	0.2 gm
$Ca(NO_3)_2 \cdot 4H_2O$	0.8 gm

Haas and Reed's A to Z Solution (includes trace elements). Weigh out the following and dissolve in 1 liter of distilled water:

H_3BO_3	0.6 gm
$MnCl_2 \cdot 4H_2O$	0.4 gm
$ZnSO_4$	0.05 gm
$Al_2(SO_4)_3$	0.05 gm
$CuSO_4 \cdot 5H_2O$	0.05 gm
$NiSO_4 \cdot 6H_2O$	0.05 gm
$Co(NO_3)_2 \cdot 6H_2O$	0.05 gm
KI	0.03 gm
LiCl	0.03 gm
KBr	0.03 gm

TiO$_2$ .... 0.03 gm
SnCl$_2 \cdot$2H$_2$O .... 0.03 gm.

The experimental group solutions should be prepared from the formula of the control solution, omitting one mineral. Containers should be carefully labeled, and seedlings should be observed periodically so that the effects of mineral deficiency, if any, can be observed.

D. TROPISMS

1) PHOTOTROPISM

Observation of classroom and outdoor plants will show the influence of light. The leaves and stems of healthy growing plants will be oriented toward the source of light. If desired, students should be asked to develop their own controlled experiments.

2) GEOTROPISM

A demonstration of positive geotropism can be set up similar to that used to observe root development in germinating seedlings (Lesson 11).

Corn seedlings should be germinated in a petri dish which contains moist blotting paper. As soon as roots begin to form, the seedlings should be transferred to a beaker which has been lined with blotting paper and filled with sand, cotton, or other supporting material (as described in Lesson 11). Four seedlings should be spaced evenly around the side of the beaker, between the glass and the blotting paper. Each seedling should be positioned so that the root of the first points downward, the root of the second points horizontally to the left, the root of the third points upward, and the root of the fourth points horizontally to the right. *How can it be determined that roots show a positive response to gravity?*

To remove the possible effects of light, a second similar set up should be prepared and left in the dark. Both set ups should be examined the following day.

Negative geotropism can be demonstrated by using mature geranium plants. Select four plants of approximately equal size. Place two of these plants in a horizontal position by placing the pots on their sides. One should be in the dark, and the other in the light. The other two plants should be left in an upright position, one placed in the dark, the other in the light. All should be watered regularly and observed periodically for 2–3 days. *How can you tell that stems are negatively geotropic?*

POSSIBLE QUIZ

1. Compare the mechanisms involved in growth movements with those in turgor movements.

2. What experimental evidence would lead you to believe that auxins are diffusable substances?
3. What are some horticultural uses of auxins?
4. Discuss the world distribution of plants in terms of photoperiodism.
5. Explain the possible relationships existing between growth movements and tropisms. Is there a similar relationship between turgor movements and tropisms?

READINGS

Butler, W.L., and R.J. Downs, "Light and Plant Development," *Scientific American*, December, 1960.

Greulach, V.A., "Plant Movements," *Scientific American*, February, 1955.

Leopold, A.C., *Auxins and Plant Growth*. Berkeley: University of California Press, 1955.

Naylor, A.W., "The Control of Flowering," *Scientific American*, May, 1952.

Salisbury, F.B., "Plant Growth Substances," *Scientific American*, April, 1957.

Schocken, V., "Plant Hormones," *Scientific American*, May, 1949.

FILMS

"Multicellular Plants, Part III: Regulation of Growth." 26 minutes, sound, color, $8.15. McGraw-Hill Book Co., Text-Film Division, 330 W. 42nd Street, New York, N.Y. 10036.

"Multicellular Plants, Part III: Seeds and Germination." 24 minutes, sound, color, $8.15. McGraw-Hill Book Co., Text-Film Division.

MODELS

1. Plant growth experiment kits and plant mineral deficiency kits may be obtained from Ward's Natural Science Establishment, Inc., P.O. Box 1712, Rochester, New York 14603.

Lesson 17

ECOLOGY AND THE TERRARIUM

Lesson time: 45 minutes
Laboratory time: 90-135 minutes

AIM
To develop a basic understanding of the principles of ecology through the use of a student developed terrarium.

MATERIALS
Battery jars, gallon jars, old aquaria, or specially purchased terraria cases; Sphagnum moss; Potting soil; Humus; Gravel; Charcoal; Coarse sand; Insectivorous plants; Animals suitable for a terrarium; Containers of water; Small cans with both ends removed; Litmus paper; Rulers; Films.

PLANNED LESSON

1. Basic principles

Ecological principles can be introduced by the use of two short films: "Plant-Animal Communities: Physical Environment" (11 minutes), and "Plant-Animal Communities: The Changing Balance of Nature" (11 minutes). These films illustrate various natural factors affecting plant-animal communities, and analyze these factors as they influence a community of living things (*biome*). As ecology deals with the interrelationships of plants and animals and their environments, and as there are so many varied plant-animal communities and environments, these films allow the teacher to explore the diversity of such interrelationships within the classroom situation. If he wishes, and as time allows, he may take his class on an ecological excursion. Vacant lots, local parks, ponds, backyards, and wooded areas are just as valuable in demonstrating ecological principles as are botanic gardens or other simulated environments.

To vividly illustrate some basic ecological principles, a classroom terrarium can be planned and developed by the teacher and students. Reinforcing the students' concept of the scientific method, they should be made aware that certain information is needed before actual planning begins. Some basic principles which should be brought in here include:

A. Communities

All plants and animals live in groups or communities. Students should be able to infer from observation and past experience that a number of members of the same species will be found in a community. Isolated or individual (single) members rarely are found in a natural setting. The community derives its name from the dominant species. For example, a maple forest contains many other species of living things, both plant and animal.

B. Special Needs

All communities have special needs. All plant and animal species have special needs for survival. These include nutrition, water, and climate. It should be apparent that those species living together in a community are able to live together because their needs for survival are similar.

C. Succession

All communities undergo succession. The principle of succession refers to a state of constant change which eventually leads to the replacement of one community by another. Many factors may influence such a change. Changes in climate, food sources, relations with other members of the community, fire, and erosion all influence succession. The teacher can select one of these factors, such as fire, and ask his students to discuss how this would affect the structure of the community and the dominance levels of its members.

D. Interdependence

The principle of interdependence refers to the equilibrium in the working relationships within the ecological community. This interdependence maintains a biological balance in nature. For example, small fish feed upon microscopic plants and animals (plankton), larger fish feed upon the smaller fish, and man feeds on both large and small fish. It might be interesting to have students propose the possible steps in other food chains or food webs. *What might be expected to occur if one member of a food chain or web were eliminated?*

E. Competition

The principle of competition refers to the struggle for survival among the members of a community. Their special needs and interdependence must be in equilibrium if the community is to be successful. All of the preceding principles are related to each other in the total concept of the *ecosystem*. The teacher should relate each of these principles, as they are discussed, to the examples presented in the introductory films.

2. Field study

An ecological excursion to a nearby park, vacant lot, yard, field, wooded area, or sand dune, can be planned by your students. They should be reminded that careful planning for any such trip is essential. Official permission must be granted by the administrators; parental permission, in writing, must be obtained; and all other plans must meet the teacher's approval. If privately owned land is used, permission must be obtained in advance. The plot to be studied must be left as it is found.

The area to be studied should be selected so that it can be conveniently visited two or more times during a one-week period.

A. Conditions

Students should observe various factors which would have an effect on the plant-animal life of the area: soil; exposure to direct sunlight; exposure to rainwater; runoff; and exposure to wind and air contaminants. Students should look for such things as:

1) windbreaks—buildings, trees, shrubs, or other structures;
2) evidence that these structures *do* serve as windbreaks;
3) the location of the structures relative to the prevailing air currents in the area. *How can you determine the direction of prevailing air currents?*
4) evidence that air contaminants such as soot and dust have settled on the plot (this may be checked by placing a piece of white paper or cloth in a protected area of the plot for several days before examining).

The amount of sunlight the plot receives will be affected by buildings, trees, shrubs, and other structures. If this area is shaded by such structures, students should estimate what percent of each day is shaded. *How would you estimate the amount of shading?*

Slopes and plant cover are major factors which affect the runoff of rainwater. *What evidence, if any, can you find of runoff? Where do you find evidence that soil has washed away?*

The soil can be examined in a variety of ways. To determine soil firmness, a sharp instrument, such as a ruler, can be pushed into the soil at different locations on the plot. Another method, of course, would involve measurement of the rate of water percolation. Students should use a small can with both ends removed and a small amount of water. Push one end of the can approximately $\frac{1}{2}$ inch into the soil, and pour a measured amount of water into it. *How many seconds does it take for the water to percolate into the soil?* Compare this rate with the rate at other localities.

Students can determine whether the soil is *acidic* or *basic* by preparing a suspension of water and soil. This should then be tested with litmus paper. Blue litmus turns red or pink in the presence of an acidic solution. Red litmus turns blue if the soil is basic.

B. Plant Life

Have students observe and list the different kinds of plants growing on this plot. *Which species seems to be dominant? How can this be determined? Why do some species of plants seem to grow better than others on this plot?*

C. Animal Life

Observe and list the different kinds of animals living on or visiting the plot. *What evidence can you observe indicating the presence of animals?*

After all data has been carefully collected and recorded, a report should be made interpreting the findings. Students should be asked to describe the ecological principles they were able to observe in operation. *Was the area studied capable of supporting an ecosystem?*

Students should be encouraged to explore the aspects of ecology in terms of conservation. The teacher should check with local museums and conservation groups for information concerning possible field trips and ecological exhibits.

PERTINENT FACTS

- Ecology is the study of the relationship of plant-animal communities to their total environment.
- Communities are characterized by special needs, succession, interdependence, and competition.

3. Preparing a terrarium

Now that your students have become familiar with some basic principles of ecology, they should be encouraged to plan and develop terraria with your guidance. Suitable containers include battery jars, wide mouth gallon jars, old aquaria, or specially purchased terraria cases. A large rectangular case would be the simplest to use. It should have a cover of glass to prevent excessive evaporation. Care should be taken not to seal this cover, in order that air may circulate freely.

Many ecological principles can be observed and studied by the development of several different kinds of terraria. Students should be encouraged to determine what factors must be controlled to maintain proper environmental conditions. Conditions will vary, depending upon the types of specimens to be kept. In general, there are four main types of terraria, classified in terms of climate conditions. These include swamp, bog, woodland, and desert habitats. The conditions required for each of the main types are:

A. Swampy

A shallow pool of water should be created in the terrarium. The soil on the side of the terrarium should be several inches above the water line. If possible, soil from a swampy area should be transported to the classroom and placed in the terrarium, on top of a base layer of gravel. An alternate method of preparation would be to use a mixture of sphagnum moss and potting soil to be placed over a layer of gravel. A layer of living or dead sphagnum should be placed over this as a top layer. Pieces of charcoal should be added to the gravel. The layer of gravel provides drainage, while the garden soil-sphagnum mixture provides an acid soil. The bits of charcoal absorb odors. After the soil is placed in the terrarium it should be thoroughly soaked.

The swampy terrarium is an excellent place in which to raise insectivorous plants. The care of these plants depends upon the species. Although all insectivorous plants require a moist or aquatic environment, there are differences. Sundews generally grow on moss-covered rocks or logs above the surface of the water, rarely in contact with it; the Venus flytrap requires a moist soil with good drainage; the pitcher plant must have at least a few of its roots in water. All require an acid soil, usually growing in sphagnum or peat moss bogs.

To protect these plants from the dry air of the heated classroom, the humidity in the terrarium should be kept high. Although some sunlight is necessary, excessive amounts will cause damage and may even kill the plants. They must be observed frequently when the terrarium is first established, to determine proper sunlight requirements. It will take approximately two weeks after planting for new plant growth to become apparent. For an interesting extension of this topic, the teacher may wish to refer to Lesson 18, Insect-Eating Plants. This lesson includes student activities centering around the swampy terrarium.

The pool of water in the terrarium might contain insect larvae and small fish or tadpoles. If large enough, this pool can be set up, like any small aquarium, with plants and fish.

B. Bog

The preparation of base and soil in a bog terrarium is the same as in a swampy terrarium. The chief difference in physical condition is the amount of water. Here, excess water should be restricted to the gravel layer which should *always* contain some water. Bog plants of various types, as well as insectivorous plants, will do well here. Animals which would be suitable include all small amphibians (newts, toads, salamanders).

C. Woodland

The most versatile terrarium is the woodland arrangement. A great number of plants and animals do well here. The preparation should include a base of gravel, sand, and humus well mixed. Regular potting soil should then be placed on this base. The size and number of animal specimens will depend upon the size of the terrarium. Care must be used in the selection and arrangement of plant specimens. Most small animals of the woodland will do well in this terrarium. They include beetles, snails, snakes, frogs, lizards, and toads.

D. Desert

This is the simplest of all types to prepare and maintain. The base should be composed of about 2 inches of coarse sand. This should then be covered with fine desert sand, which can be purchased from most pet shops or commercial supply houses. The base layer should be moistened, but the top layer should be kept dry. A small pan for water should be embedded in the sand so that it is level with the sand surface. Cactus plants should now be planted in a pleasing arrangement. The area around the plants should be lightly watered when planting is complete. A few attractive rocks and a piece of driftwood might be included if there is adequate space. Some snakes do well in a desert terrarium, as do certain lizards. Other desert creatures might be used instead. Since humidity should not be high, no cover of glass is needed. A piece of wire mesh over the top will be enough to keep in the animals. Make certain that the mesh is secured so that it cannot be pushed aside by the animals.

Students must pay special attention to the interrelationships among all the specimens when planning the terraria. As proper soil, water, and temperature will generally insure satisfactory plant growth, this is perhaps a simpler problem than that of the animals in the terraria. Students should be aware of the differences in feeding habits and food requirements of the animals they select. It should be noted that some animals will not feed normally in captivity until they have become accustomed to their surroundings. Techniques of forced feeding may be required. On the other hand, cold-blooded animals may survive for long periods of time without feeding. Students will need to research this information carefully before making final decisions.

PERTINENT FACTS

- The use of classroom terraria brings the outdoors in and provides a microclime for classroom study.
- Preparation of the soil bed in a terrarium is important to its success.
- Communities, successions, interdependence, special needs, and dominance, can be studied in great detail in terraria.

POSSIBLE QUIZ

1. Discuss the ecological principles functioning in an ecosystem.
2. Fully explain the consequences of a major flood on an ecosystem.
3. Outline the steps which would be followed in studying the ecology of a plot of land.
4. Define microclime, ecosystem, and biome.
5. Select one type of terrarium and explain how it would be developed. What ecological principles can be observed in operation in it?

READINGS

Daubenmire, R.F., *Plants and Environment*. New York: John Wiley and Sons, Inc., 1959.

Dice, L.R., *Natural Communities*. Ann Arbor, Michigan: University of Michigan Press, 1952.

Hanson, H.C., and E.D. Churchill, *The Plant Community*. New York: Reinhold Publishing Corp., 1961.

Polunin, N., *Introduction to Plant Geography*. New York: McGraw-Hill Book Co., 1960.

Turtox Service Leaflet, *The School Terrarium*. Chicago, Illinois: General Biological Supply House, Inc., 1959.

FILMS

"Chain of Life." 11 minutes, sound, color, $3.40. Pictura Films, 41 Union Square W., New York, New York 10003.

"Plant-Animal Communities: The Changing Balance of Nature." 11 minutes, sound, color, $3.90; bw, $2.15. Coronet Films, Coronet Building, Chicago, Illinois 60601.

"Plant-Animal Communities: Physical Environment." 11 minutes, sound, color, $3.90; bw, $2.15. Coronet Films.

MODELS

1. Plants and soils can be obtained from local flower shops and pet stores as well as from Carolina Biological Supply Company, Burlington, North Carolina 27215.
2. Complete kits for the development of each type of terrarium can be obtained from Carolina Biological Supply Company.

Lesson 18

INSECT-EATING PLANTS

Lesson-laboratory time: 90 minutes

AIM

To study in some detail the cultivation and behavior of insectivorous plants, with emphasis on their unique responses to stimuli.

MATERIALS

Terrarium case; Sphagnum moss; Potting soil; Humus; Gravel; Charcoal; Litmus paper; Insectivorous plants—Venus flytrap, Sundew, Pitcher plant.

PLANNED LESSON

1. Cultivation

The cultivation of insectivorous plants is a unique and enjoyable experience. Information concerning these plants is generally not included in depth in textbooks, so that it may become necessary for your students to do outside reading.

The observation of insectivorous plants will illustrate the principle of adaptation to the environment. Although these plants are not very common, they are found throughout the world in tropical and temperate regions. Several interesting types of insectivorous plants are found in the United States. These include the pitcher plant, with a leaf cavity filled

INSECT-EATING PLANTS 143

with water; the sundew, with its series of dew-tipped hairs; and the Venus flytrap, with its jaw-like leaves. In accordance with the information in Lesson 17, these plants should be cultivated in a swamp or bog terrarium.

2. Characteristics

From their outside library research, students should have developed a basic understanding of the characteristics of these insect-eating plants:

A. Pitcher Plant (Sarracenia purpurea)

This plant is so named because its leaves resemble pitchers. These structures are hollow cylinders open, at least partially, at or near the top, as illustrated in Figure 18-1. Inside this cylinder are long spines which grow along the rim. These spines grow downward and prevent trapped insects from crawling out. Rainwater and digestive juices partially fill the lower quarter of the "pitcher." Insects are attracted to these plants by the secretions of numerous glands lining the pitcher opening. Once the insect enters the mouth of the plant it becomes trapped inside by the spines which point downward. After a period of time, the insect becomes exhausted through repeated attempts to escape, falls into the fluid at the bottom of the pitcher, and drowns. The digestive enzymes in the water decompose the body of the insect, and the end products of this digestion are absorbed directly into the cells lining the cavity. An interesting phenomenon to be noted here is that certain insects which feed on decaying organic matter will be lured to their deaths by the odor of decomposing insects.

B. Sundew (Drosera rotundifolia)

As illustrated in Figure 18-2, the leaves of this plant resemble flat discs, and are supported by elongated filamentous petioles. Bristles or hairs can be seen to radiate from the edges and upper surface of the disc. The tips of these bristles produce sticky secretions of digestive juices which serve to trap and digest insects. When an insect lands on a leaf, it becomes covered with the sticky secretion; at the same time, the bristles begin bending inwardly, covering the insect. While the insect is trapped by the bristles the glands at their tips secrete large quantities of digestive juice. Several days later, when absorption has been completed, the bristles return to their original position.

C. Venus Flytrap (Dionaea muscipula)

As can be observed in Figure 18-3, the leaves of the Venus flytrap resemble jaws or an interlocking trap. The outer edge of the leaf is fringed

Figure 18–1. Pitcher Plant.
Courtesy Carolina Biological Supply Company

Figure 18–2. Sundew.
Courtesy Carolina Biological Supply Company

Figure 18-3. Venus Flytrap.
Courtesy Carolina Biological Supply Company

with sharp spines, and the leaves are supported by a flat petiole. The upper surface of the leaves are usually bright red in color, the color being the result of the presence of hundreds of digestive glands. A trigger-like device composed of three bristle-like hairs is found on each half of the leaf. When an insect alights on a leaf, the triggering mechanism will cause the leaf to snap shut if two of the three hairs are touched, or if one hair is touched twice. The two halves of the leaf will snap shut, and the sharp spines will interlock, trapping the insect. *What happens now?* Digestion takes about 14 days, after which time the leaves will reopen.

During this lesson, students should be encouraged to consider the plant's need for animal tissue. Since it will not be obvious to the students, the teacher may have to explain that swampy or bog soil is usually deficient in nitrogen-fixing bacteria and that the root systems of insectivorous plants are poorly developed, making nitrogen absorption difficult. *What other source of nitrogen do these plants have?*

3. Activities

A. PITCHER PLANT

Students may catch flies or other insects and feed the pitcher plant. Students should be encouraged to observe the plant daily. Hopefully, they will observe a live insect entering the plant. The contents of the pitcher may be tested with litmus paper. *Is it acidic or basic? Is undigested material eliminated? How?*

B. Sundew

Observation and feeding can be accomplished in the same manner as mentioned for the pitcher plant. *Does the sundew have a preference for different types of insects? How was this determined? How can you determine the degree of sensitivity of this reaction? Will inanimate objects produce a reaction?*

C. Venus flytrap

Observation and feeding should be performed again here. Have your students identify the mechanism by which the trap is sprung. *Will the trap accept undigestible food or inanimate objects?* (It should be noted that bits of fresh meat can be used as a food material.) The teacher should explain that rejection will require approximately 24 hours. Students should not jump to conclusions if a trap does not open immediately after an undigestible particle has been trapped inside.

A simple experiment may be devised to determine whether constant stimulation affects plant responses. Two Venus flytraps should be used. One should be stimulated repeatedly, while the other is not. All other conditions should be kept the same. *Why is one plant left alone? What role does this plant play in the experiment? Why is it important? Would it be better to use four plants, two to be stimulated and two not? Why? What happens if too much food is placed within the trap?*

PERTINENT FACTS

- The leaves of insectivorous plants function in a manner similar to the human stomach, when they are actively secreting enzymes.
- Insect ingestion is related to the plants' need for nitrogen.
- Each species of insectivorous plant has a distinct method of trapping insects.
- The response mechanism of some insectivorous plants is the result of a change in turgor pressure. (Lesson 16.)

POSSIBLE QUIZ

1. What are the basic similarities in the food-getting mechanisms of any two insectivorous plants? Differences?
2. How are the leaves of the pitcher plant, sundew, and Venus flytrap adapted for food getting?
3. What will be the effects of over-feeding on insect-eating plants?
4. Do insectivorous plants raised in the classroom require animal tissue ingestion for survival?

5. In terms of the material previously learned about turgor pressure, give a possible explanation of what happens within the plant to cause trapping.

READINGS

Baker, B., "Carnivorous Plants: Exploring Their Special Ways of Survival," *Professional Growth For Teachers, Science,* Croft Educational Services, December, 1965.

Lloyd, F. E., *The Carnivorous Plants.* New York: Ronald Press, 1963.

Poole, L., and G. Poole, *Insect-Eating Plants.* New York: Thomas Y. Crowell Co., 1963.

Turtox Service Leaflet, *Insectivorous Plants.* Chicago, Illinois: General Biological Supply House, Inc., 1959.

FILMS

"Living Traps." 10 minutes, sound, color, $3.65. Sterling Movies, U.S.A., Inc., 43 W. 61st Street, New York, N.Y.

"Plant Oddities." 10 minutes, sound, color, $3.40. John Ott Pictures, Inc., P.O. Box 158, Lake Bluff, Illinois.

MODELS

Living insectivorous plants, plastomounts, and preserved specimens may be obtained from Carolina Biological Supply Company, Burlington, North Carolina 27215; and from General Biological Supply House, Inc., 8200 South Hoyne Avenue, Chicago, Illinois 60620.

Unit IV

THE ORGAN SYSTEMS OF MAN

Lesson 19

LOCOMOTION AND SUPPORT

Lesson time: 135 minutes
Laboratory Time: 90 minutes

AIM

To develop an understanding of the structure and function of the skeletal and muscular systems, and their interrelationships.

MATERIALS

Microscopes; Prepared slides of bone and muscle tissue; Human skull; Articulated and disarticulated skeletons; Fresh long bones; Preserved frogs; Dissecting trays, pins, needles, scalpels, scissors, tongs; Forceps or pliers; Rubber tubing; Containers for boiling water; Household detergent and bleach.

PLANNED LESSON

1. Bone anatomy

The teacher may begin with a study of the structure of bone. If notified several days in advance, a local butcher or food market can supply a complete long bone for classroom use. You should request that it be cut longitudinally so that the internal structure can be shown. As structures are pointed out and discussed, a diagram similar to Figure 19-1 can be placed on the chalkboard.

If available, disarticulated bones from a skeleton may be distributed so that each table will have one for the students to observe. Pieces of old skeletons which are to be replaced should be stored in cardboard boxes and kept for this purpose. If these are not available, chicken bones may be used. On larger bones the teacher will be able to point out the grooves over which blood vessels and nerves pass. This type of structure can be graphically illustrated by using a skull to point out the holes (*foramena*) on either side of the lower jaw, through which blood vessels and nerves pass. On many skulls students will be able to see the groove on the edge of the jaw. This may help them to understand the idea of the "glass jaw" of boxers. When the nerve is pinched against the bone, the person will lose consciousness. *How does this affect someone, like a boxer, who has no groove or only a very shallow one?*

A simple activity can be used to demonstrate this same principle as it applies to the ulnar nerve. By bending the elbow so that a 90° angle is made between the forearm and upper arm, the students will be able to feel the groove on the inner surface of the elbow, which is the location of the ulnar nerve. They should be directed first to locate the bump of bone on the inner surface of the elbow. This will be approximately three-quarters of an inch in size. Then by placing a finger in the groove it produces, moving the arm slightly forward, and moving the finger back into the groove, the ulnar nerve will be pinched against the bone, producing the typical "funny bone" sensation.

If the long bone obtained from the butcher is fresh, the class will be able to identify the blood vessel entering the central area or shaft in which the bone marrow is found. The teacher might briefly point out that the marrow is blood-producing tissue. The outer shiny coating, or *periosteum*, can also be shown.

If the teacher feels it will be worthwhile, the microscopic structure of bone can be presented here. A diagram similar to Figure 19-2 can be used to point out the Haversian canal system and its structure. The role of each part may be covered by indicating the location of lacunae, osteocytes, and lamellae. *What relationship exists between the location of the blood vessel of each system and the horizontal position of the canaliculi? How does bone grow in diameter? How does this affect the size of the shaft?* If available, slides of bone tissue may be studied in the laboratory.

2. Organization of the skeletal system

Starting with the skull, it would be worthwhile to point out the major

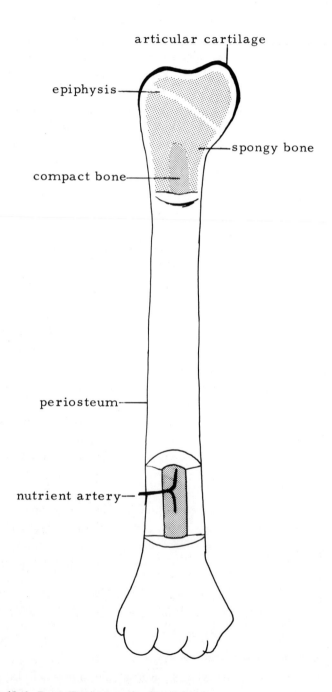

Figure 19–1. Long Bone, Longitudinal Section.

LOCOMOTION AND SUPPORT   153

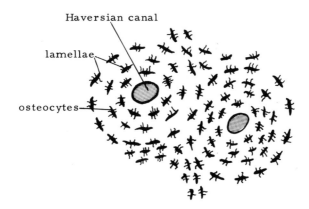

Figure 19-2. Haversian Canal System, Microscopic Structure of Bone.

bones of the skeletal system. The teacher can use a human skull or a plastic model of one, and pass out diagrams similar to Figure 19-3 to identify the cranial bones for the class. Special mention should be made of the *sutures* which can be seen as the borders of the bones. *Why is the dove tail arrangement of these margins better than a straight line closing? What is the* soft spot *in a baby's skull?*

Other special points which might be mentioned include the foramen magnum, or "big hole" at the base of the cranium through which the spinal cord passes in the human; the hinge joint of the jaws; the ear canals in the temporal bones; and the bump or occipital protuberance, on the back of the skull, to which the broad muscles of the back attach.

If available, a skull which has a removable top can be used to show the inner surface of the cranial bones. The "Turk saddle," or *sella turcica*, can be easily identified from its appearance, and will be observed at the floor of the cranium. From its name and appearance students should be able to see that it acts as the "saddle" for some structure. The teacher may want to identify the structure at this point simply by explaining that it is a gland called the pituitary.

If several skulls are available and are of the open type, students may notice that the thickness of the occipital bone may vary markedly. This racial characteristic sometimes helps to identify the skull of a Negro (thicker bone), as opposed to a Caucausian (thinner bone). Although students may be interested in the reasons for this, no explanation can be given other than that it is a racial characteristic, as is eye slant or hair type.

Many of the external features observed on the skull can be felt by students palpating their own heads. Chewing movements will help them to understand the workings of the hinge joint of the jaws if they place

154 LOCOMOTION AND SUPPORT

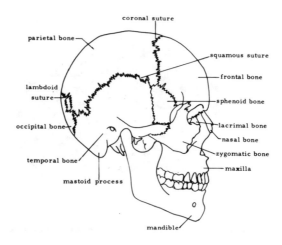

Figure 19-3. Cranium.

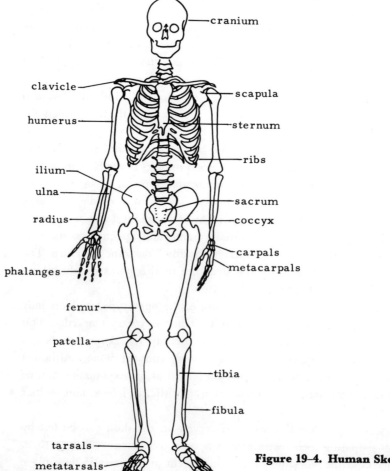

Figure 19-4. Human Skeleton.

their fingertips on their cheek, over the joint. Some students may hear a clicking noise in the joint. The teacher might point out that some very new research indicates that habitual gum chewers may develop this characteristic clicking, as well as serious wear at the joint and related aching.

Moving on to the rest of the skeleton, students can use a duplicated sketch, as of Figure 19-4, to label, as the teacher points out the bones of the skeleton. As the teacher does so, students should attempt to palpate the bones of their own bodies.

The first vertebrae, the atlas and axis, can usually be seen on the skeleton without difficulty. *Why is the atlas so named? The axis? What movement of the cranium do they provide?* Students should be aware that the skull, together with the vertebral column and bones of the thorax, compose the axial skeleton. The vertebral column can be divided into sections. These are the cervical vertebrae which have prominent projections. By bending their necks forward, students will be able to feel the protrusions of these vertebrae along the back of the neck. The atlas and axis are the first two cervical vertebrae. The thoracic vertebrae may be pointed out by the teacher on the skeleton or a chart. *What other structures are attached to these?* The lumbar vertebrae can now be indicated. *Where might the illness known as lumbago be centered?* Students should pay close attention to the sacrum so that they can identify the five vertebrae which have been fused to form this structure. Usually the coccyx will be attached to the skeleton by means of a small wire. Students can see that several bones are fused.

The synthetic material which is used in articulated skeletons to represent the cartilage of the skeletal system should be pointed out. This cushioning material should be identified by the students with the aid of reference materials, and its function should be brought out. *Why does a slipped disc of cartilage cause pain?* The various curvatures of the spine should be observed and their function discussed. Abnormal curvatures, such as *scoliosis* (lateral curvature) and *kyphosis* (hunchback) should be mentioned. Perhaps student reports can be given on these and other bone diseases.

Moving around to the front of the skeleton, the third group of axial bones can be seen. Known as the bones of the thorax, these are the sternum and ribs. Students should be asked to identify the seven pairs of *true* ribs and the five pairs of *false* ribs. *Why are the last five pairs called* **false** *ribs? Why are the last two pairs called* **floating** *ribs?* The sternum or breastbone can be observed at the same time. *What function does the rib cage appear to have?*

Students can easily feel their clavicle or collar bone, by bringing their shoulders up and forward. *Why does the clavicle break if you fall hard against the shoulder?* The shoulder blade, or scapula, can be made more prominent by bringing the hand up behind the same side of the body, so that the back of the hand touches the vertebrae as high up the spinal column as possible. The humerus of the upper arm presents no problem in palpation. An articulated skeleton should be used to point out the location of the radius and ulna of the lower arm. By holding their arms out so that the back of the hand touches the table surface, the students will have placed their own radius and ulna in a position so that they are parallel (the radius or thumb-side bone should be outermost). By keeping the arm in the same position, and bending at the wrist so that the palm is now in contact with the table surface, the students will have caused the characteristic rotation of the lower arm. *What is the position of each bone now?* They will be able to see this more clearly if the same maneuver is now repeated using the articulated skeleton. The bones of the wrist (carpals) can now be shown on the skeleton, as well as those of the fingers (metacarpals, phalanges), which can also be palpated. Most of the time, a broken "wrist" is actually a break in a bone of the lower arm. *Why would pressure on the wrist cause the break to occur in the lower arm?* Young bones tend to break only part-way through. *Why is this called a **greenstick** break? Which is the largest and longest bone of the upper appendages?*

Moving on to the lower portion of the skeleton, students will see that the pelvis is composed of several bones which have fused together: the ilium, ischium, and pubis. As students are usually interested in determining the sex of a skeleton, a study of the structure of the pelvis will be helpful. They will see that the female pelvis tends to be wider than that of the male, and that it tips forward somewhat more than in the male. As can be observed, these are relative factors which require the comparison of several skeletons, and will also depend upon the age of the skeleton. The small piece of cartilage, the symphysis pubis, seen at the front joining of the pelvis, should be pointed out and its importance mentioned here. The need for expansion during childbirth can be understood and may be related to another characteristic of the female pelvis, the larger pelvic opening.

The similarities between the upper and lower limbs can now be shown: the femur and humerus, tibia and fibula with radius and ulna. The differences should also be mentioned, including the limited movement at the hip joint as compared with the movement at the shoulder. *How does movement at the ankle compare with that at the wrist?* The patella or knee cap should also be pointed out. *What is its function?*

3. Types of joints and the movement they permit

As a unifying study, the teacher can now present the types of movement which occur at the joints previously studied. This might be done by selecting a joint and then having students move the bones composing that joint, to determine what movements are possible. Some of the joints which should be mentioned include synovial, hinge, spheroidal, and sutures of the skull. Types of movements at synovial joints which should be brought out include flexion, extension, abduction, adduction, protraction, and retraction.

4. Disorders

Students are frequently interested in various illnesses and disorders of the systems of man. This study provides an excellent opportunity for the teacher to dispell many half-truths and tales students may have heard, as well as to encourage positive attitudes toward medical treatment, and quick and proper care when needed. Some of the more common questions students have frequently involve arthritis (arthro-joint, itis-inflammation); bursitis, or inflammation of fluid-filled sacs (bursae act as cushions in certain joints); dislocation, in which parts of a joint are pulled or pushed out of position; and bunion, or the swelling of a bursa of the foot. *How does arthritis differ from bursitis?*

5. Skeletal dissections and preparations

Any bony skeleton can be used. For comparative study, laboratory specimens previously used, such as the frog, can be dissected carefully so that the skeleton is exposed. (Students can also bring in chicken bones from home.) Students should try to observe the relationship between the tendons, ligaments, and bones, as well as their connections. They should compare the skeletons of lower animals to the skeleton of man.

In the preparation of a skeleton, the student should make certain that as much tissue as possible has been removed from the bone, by gingerly scraping with a dull knife or other appropriate tool. These bones should then be placed in a container of boiling water to which a small amount of household detergent has been added. The bones should be kept in the boiling water for approximately five minutes. Using forceps or tongs, the bones should be removed at the end of this time, and any clinging tissue should be gently brushed off with a soft toothbrush. If some tissue is not easily removed, that bone should be returned to the boiling water for an additional three minutes.

The bones may be bleached by using cleansing powder, such as Ajax or Comet, and scrubbing with a wet toothbrush; or they may be soaked in a dilute solution of household bleach or sodium hypochlorite for one to two hours. The bones should then be rinsed in tap water and allowed to dry.

A worthwhile project would be to have students articulate these bones. If the skeleton is small, the bones can be glued or otherwise attached to a flat surface. The bones of larger skeletons may be wired into their natural positions by using large dissecting pins and a small hammer to make holes in the bone. Fine wire, such as piano wire, and needle-nosed pliers or forceps will also be needed.

6. Muscle anatomy

The various types of muscle, and their relationships to the skeletal system should be introduced here. The function of muscle—its contractility—should be stressed. *Which muscles would probably be identified as voluntary? Involuntary? Cardiac?*

A brief study of each type in terms of microscopic structure, using prepared slides or charts, can be introduced with emphasis on the voluntary muscles, since these are interrelated with bone movement. In striated muscle students should be familiar with fiber shape and nuclear arrangement. They should see that the cytoplasm of each cell or fiber contains fine fibers or *myofibrils*. *What is the function of the myofibrils within the cytoplasm? Can alternating light and dark areas be seen in the cytoplasm? How does shape and nuclear arrangement in smooth muscle differ from that in striated muscle? Are light and dark bands visible in smooth muscle?*

In cardiac muscle students should observe the dark-light banding (intercalated discs) of the myofibrils. *What is the function of these discs? In what ways is cardiac muscle similar to smooth muscle? Similar to striated muscle?* Students should be aware that all three muscle types are supplied with nerve endings which carry messages from the central nervous system. *What role does the nervous system play in muscle innervation? What is the neuromuscular junction or motor end plate?*

7. Theories of muscle contraction

The teacher might present some of the current hypotheses concerning muscle contraction, or students may prepare reports based upon their research in scientific periodicals and texts. The prominent Huxley theory of sliding filaments should be included. The nature of the myofibrils of skeletal muscle can be considered in greater depth at this time. *According to Huxley, what causes the dark and light band arrangement?* Students must include the explanation that the myofibrils of muscle are composed of thick and thin filaments which are macromolecules known as *actin* and *myosin*. They should also include in their report an explanation of the sliding mechanism and the possible existence of cross-bridges.

8. Characteristics of muscle

Before proceeding with a discussion of specific muscles and their attachments, the teacher can develop the concepts of the all-or-none law, muscle fatigue, and muscle tonus.

A. ALL-OR-NONE LAW

Students should understand that an individual muscle fiber will either contract to its fullest extent when stimulated, or it will not contract at all. It should be stressed that the same muscle fiber will not necessarily contract to the same extent everytime it is stimulated, since various factors, including fatigue, can affect the strength of contraction. *If a fatigued muscle fiber is stimulated and it contracts to its fullest extent in accordance to the all-or-none law, how will this contraction compare with a contraction in the same muscle fiber when it is completely rested?*

B. MUSCLE FATIGUE

The concept that a muscle fiber will respond with less "strength" due to the accumulation of metabolic waste products should be developed. The idea that this accumulation of metabolic wastes causes a delay between the moment of stimulation and the actual response of the muscle fiber, should be brought out. This will provide your students with another example of the effects of various conditions on the all-or-none law. The teacher can now correlate the idea of "overtiredness" in humans with the characteristic lack of total relaxation in fatigued muscles. *In terms of the accumulation of metabolic wastes in muscle, what purpose is served by hot showers and massage immediately after strenuous physical activity?*

C. MUSCLE TONUS

The teacher can conclude the study of muscle characteristics with a discussion of muscle tone or tonus in skeletal muscle. Nervous stimulation keeps some fibers of each skeletal muscle in a state of contraction. The students should clearly understand that, in accordance with the all-or-none law, individual muscle fibers always contract to their maximum when stimulated; however, a muscle is composed of thousands of muscle fibers, not all of which are necessarily stimulated at the same time. With this in mind, students should understand that a muscle may be partially contracted, as in muscle tonus.

9. Muscles and their attachment

Before beginning a study of the major muscles of the body, some understanding of tendons and ligaments should be developed. The attachment of muscle to bone by tendons, and bone to bone by ligaments can be demonstrated at this time. Palpation of the Achilles tendon at the back of the heel can be used as an illustration. By having a student grasp the area immediately behind his ankle, and moving his foot up and down from the ankle, the tough band of the Achilles tendon can be felt.

Using a skeleton and student movements during exercise and other activities, the teacher can demonstrate the attachment of some of the major muscles, and the movements they allow. By flexing their arms, students can make the biceps muscles prominent. This familiar maneuver will provide a good point at which to begin. The location of the muscle can be shown on the skeleton. Its attachment to the humerus and radius will help to illustrate how movement of the forearm occurs. Now, holding their arms in the flexed position, students should try to explain the mechanism needed to extend the arm. The pulling of bone by muscles should be stressed at this point. The need for a muscle which will cause the opposite movement can be seen from the activity. By moving the arm from the flexed to the extended position several times, they may be able to explain the need for a muscle attached on the upper arm and below the elbow, on the side opposite to the biceps. From this activity, the concepts of pairs of muscles can be introduced and explained, as well as the pulling action of muscles. The terms "prime mover" and "antagonist" can also be introduced at this point.

Using human anatomy charts the teacher may point out some major muscles of the body and briefly mention their functions. If he prefers, he can provide students with a list of these, and have them use charts, the skeleton, and reference books to identify the location (origin and insertion) of the listed muscles and the movements they allow. Included in this list might be: trapezius, latissimus dorsi, pectoralis major and minor, biceps, triceps, deltoid, quadriceps femoris, hamstrings, gluteus maximus and minimus, gastrocnemius, external oblique, diaphragm, orbicularis oculi, orbicularis oris.

As a review and reinforcement, the teacher might relate the movement of any of these muscles to the theory of muscle contraction on the cellular level. It might be a good idea to bring in the all-or-none law here, together with the action of the actin and myosin filaments.

10. Muscle fatigue activity

A number of activities may be used as time permits. Among these are:

A. A subject places his elbow on a table, as he sits comfortably on a chair. A book should be held in that hand and then lifted up and back down to the surface of the table in such a manner that the elbow always remains on the table. The book should be brought up so that it is perpendicular to the table. This movement (flexing, extending) should be continued at a rapid pace until the subject can no longer go on. *How long did it take for the muscle to become fatigued?* Repeat after the subject is thoroughly rested, but this time use rubber tubing to make a tourniquet on the upper arm. This need not be tight, but rather just firmly tied. An alternate method would be to use a sphygmomanometer (blood pressure cuff); this can be placed around the arm and inflated. In each case, the teacher must watch carefully and see that the procedures are carefully followed. *How long does it take for fatigue to occur? What effect does reduced blood circulation have? Why must the same subject be used and why must he be well rested?*

B. The preceding activity may be performed in another manner. Instead of using a book, students may simply flex and extend the fingers of the hand as quickly as possible until fatigued. After rest, the test may be repeated with reduced circulation, as in the preceeding activity.

C. Isometric exercises might be demonstrated by an interested student. The theory of muscular contraction against an immovable object can be discussed. *Is energy used by a muscle which does not cause movement even though it is contracting? How can isometrics be of value to people who do not take active part in athletics or other forms of physical exercise?*

11. Muscle diseases and disorders

Student reports on various diseases and disorders can be used to supplement the lesson. Topics of interest can include hernia, "Charley horse," stiff neck, and paralysis.

PERTINENT FACTS

- The Haversian canal system is the microscopic unit of bone structure.
- During bone growth, cartilage is slowly replaced with bone tissue.
- Man has an ossified skeleton which may be divided into two parts, the axial and appendicular skeletons.

- The skeletal and muscular systems are connected by means of tendons and ligaments.
- One prominent theory of muscle contraction involves the sliding of actin and myosin filaments.
- The all-or-none-law is the theory which explains the nature of muscle fiber contraction.

POSSIBLE QUIZ

1. With the aid of a diagram, indicate the microscopic structure of bone and explain the function of the parts.
2. List the major parts of the axial and appendicular skeletons.
3. List three kinds of joints, and indicate the kinds of motion they allow.
4. Explain and compare the three types of muscle tissue. What is the Huxley theory of muscle contraction?
5. Why does the all-or-none law hold for individual muscle fibers but not for a whole muscle?

READINGS

Du Brul, E. L., *Biomechanics of the Body*, BSCS Pamphlet, No. 5. Boston: D. C. Heath and Co., 1963.

Griffin, D., *Animal Structure and Function*. New York: Holt, Rinehart, and Winston, 1962.

Hayashi, T., and G. A. W. Boehm, "Artifical Muscle," *Scientific American*, December, 1952.

Huxley, H. E., "The Contraction of Muscle," *Scientific American*, May, 1958.

———, "The Mechanism of Muscular Contraction," *Scientific American*, December, 1965.

Katchalsky, A., and S. Lifson, "Muscle as a Machine," *Scientific American*, March, 1954.

McLean, F. C., "Bone," *Scientific American*, February, 1955.

Szent-Gyorgyi, A., *The Chemistry of Muscular Contraction*. New York: Academic Press, 1951.

FILMS

"The Human Body: Muscular System." 13 minutes, sound, color, $4.40. Coronet Films, Coronet Building, Chicago, Illinois 60601.

"The Human Body: Skeleton." 10 minutes, sound, color, $3.40; bw, $2.15. Coronet Films.

"Human Skeleton." 12 minutes, bw, $2.15. United World Films, Inc., 221 Park Avenue S., New York, N.Y. 10003.

"The Multicellular Animal, Part IV: Muscles." 30 minutes, sound, color, $8.15. McGraw-Hill Book Co., Text-Film Division, 330 W. 42nd Street, New York, N.Y. 10036.

"Muscular System." 11 minutes, sound, bw, $2.15. United World Films, Inc.

"Skeleton." 13 minutes, sound, bw, $2.15. Encyclopedia Britannica Films, Inc., 1150 Wilmette Avenue, Wilmette, Illinois 60091.

MODELS

1. Human anatomy charts, models, articulated and disarticulated human skeletons can be obtained from General Biological Supply House, Inc., 8200 South Hoyne Avenue, Chicago, Illinois 60620.
2. Plastic assembly models and kits for student use, including plastic human skull kit and miniature human skeleton, may be obtained from General Biological Supply House, Inc.

Lesson 20

THE DIGESTIVE SYSTEM AND ITS FUNCTIONS

Lesson time: 90 minutes
Laboratory time: 45–90 minutes

AIM

To present the major concepts of the structure and function of the digestive system by a study of how foods in man are physically and chemically broken down and absorbed, and how solid wastes are eliminated.

MATERIALS

Benedict's solution; Salt and sugar solutions; Pancreatin solution; Pepsin extract; Vinegar; Lemon juice; Vanilla or almond extract; Quinine cold tablet; Oil; Liquid soap; Hydrion paper; Test tubes; Eye droppers; Hollow glass tube; Apple, potato, onion; Soda cracker; Films.

PLANNED LESSON

1. Anatomy of the digestive tract

By briefly describing the nature of the tube-within-a-tube body plan as it first appears in the *annelids*, the teacher can then present the structure

THE DIGESTIVE SYSTEM AND ITS FUNCTON

of the human digestive system. A simple diagram, such as Figure 20-1, can best be used to illustrate basic anatomy. As the lesson progresses, the functions of each part of the system can simply be listed next to the appropriate place on the diagram.

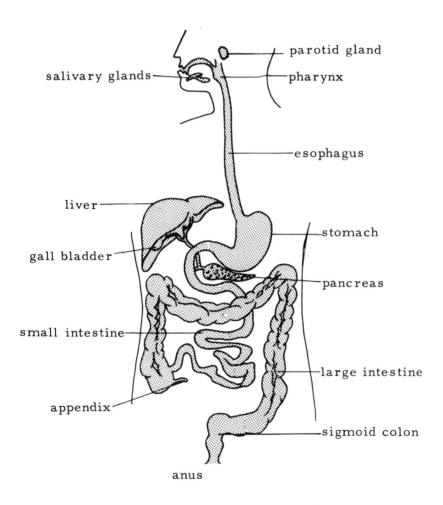

Figure 20-1. Human Digestive System.

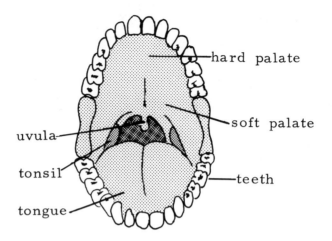

Figure 20-2. Mouth.

A. Mouth

A diagram similar to Figure 20-2 can be used to illustrate the arrangement of the teeth, tongue, uvula, and tonsils. Discussion should include the *parotid, submaxillary,* and *sublingual* glands as the three sets of salivary glands, and the role they play in saliva production and secretion. *Where are these three sets of glands located? What function do the teeth serve?* During a laboratory period students can be given time to use reference materials to determine the number and kinds of teeth, and the function of each type. If time permits, the structure and origin of a single tooth may be presented using an illustration similar to Figure 20-3.

An interesting laboratory activity can be performed to study the taste sensitivity of the tongue. A student volunteer should be selected as a subject, and asked to close his eyes tightly and to hold his nose. To test for sensitivity, clean eye droppers can be used to place various solutions on his tongue. It is essential, after one or two drops of solution have been placed on the tongue and identified as to kind of taste by the subject, that the subject rinse his mouth with water before beginning another test. A "map" of the tongue can thus be developed by the class, showing areas of the tongue and their sensitivities. Fluids which may be used include salt solution, sugar solution, vinegar or lemon juice, vanilla or almond extract, or any others of strongly salt, sweet, bitter, or sour taste. *It is essential to use only non-harmful substances,* such as food materials, for these tests. However, for a particularly bitter taste, a quinine cold tablet can be pulverized and dissolved in an eight ounce glass of water. *What functions does the tongue serve?* The pushing of a food mass into the back of the mouth by the tongue can be more easily understood if students are asked to concentrate on mouth movements as they eat something, such as a small piece of fruit.

B. Pharynx (throat)

The area at the back of the throat can be pointed out, particularly in relation to hiccuping, choking, and other phenomena in which the nearness of the esophageal and tracheal openings is important. Students should be led to understand that some means of preventing the entry of food and liquids from the common pharynx into the trachea must be present. The concept that muscular contractions regulate the tracheal opening as the food mass or *bolus* is swallowed, should be included.

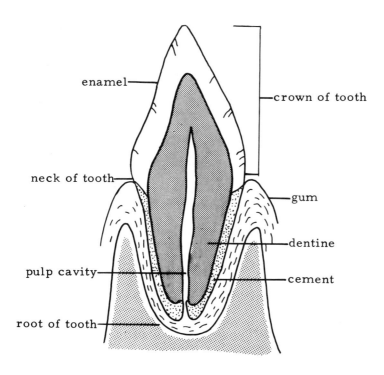

Figure 20-3. Tooth.

C. Esophagus

Muscular contractions of the walls of the esophagus and the production of small quantities of mucus should be mentioned, to indicate the process of food passage from the esophagus to the stomach. *Although the esophagus is composed of striated muscle, does an individual have voluntary control over it?*

D. Stomach

If at all possible, it would be worthwhile to illustrate the muscular contractions of the stomach, *peristalsis*, by using a film such as "The Human Body: Digestive System." This churning movement can be vividly seen. It should be pointed out that circular *sphincter* muscles constrict at the upper and lower ends of the stomach, preventing food from leaving until the proper stage of digestion has been reached. Some discussion of the rough, folded *rugae* on the inner surface of the stomach should be included to give a more complete understanding of the mechanical process of churning, and its efficiency. The presence of many glands can also be mentioned. Their functions might better be discussed in relation to the chemical process of digestion.

E. Small Intestine

It should be made clear that the intestines are named not on the basis of their overall length, but on the basis of their diameter. The *duodenum*, *jejunum*, and *ileum*, as the three parts of the small intestine, can be included. The presence of tiny projections, *villi*, can be shown on a chart or diagram.

F. Large Intestine

What relationship does this structure have to the small intestine in terms of size? Using Figure 20-1, the *cecum;* ascending, transverse, and descending portions of the *colon;* and the *rectum*, as parts of the large intestine, can be illustrated. Mention should be made of the appendix as a small finger-like outpocketing of the cecum. *What is its function?*

G. Accessory Organs

Although not part of the tube through which food moves, the teacher may mention several accessory glands which play a major role in the digestive process. Students should be aware that the liver, located under the diaphragm; the gall bladder, which is attached to the undersurface of the liver, and is connected to the duodenum by the bile duct; and the pancreas, which lies just below the stomach, function in the chemical aspects of digestion.

2. Physiology of the digestive system

Using Figure 20-1, the teacher can begin listing foods, enzymes, and other digestive materials next to each organ. By doing this in an orderly arrangement for protein, carbohydrate, and fats, the teacher can show the degradation of complex molecules to the end products of digestion, and the use of these end products in the cells.

A. Foods

The inorganic salts found in the diet should be mentioned. Included might be sodium chloride, potassium, iron, phosphate, and iodide ions. All of these can be studied as the functioning of the various systems is studied. Carbohydrates, proteins, and fats, the organic foods, might be presented in greater detail as they will be followed throughout their chemical breakdown.

1) Carbohydrates

Students will need to understand that carbohydrate molecules are combinations of simple sugars or *monosaccharides*, and that the digestion of carbohydrates is a process which will reduce polysaccharides to disaccharides to monosaccharides. The idea that glucose is the monosaccharide found in the blood will help students understand that this is the monosaccharide most widely used by cells to produce energy.

2) Proteins

It will be necessary to point out that the structural units of proteins are amino acids. *What will the end products of protein digestion be?* Here again the teacher may include material covered earlier in the term in the study of the cell, by pointing out that amino acids are used for production of all structural components.

3) Fats

The teacher can simply list the end products as fatty acids and glycerol. *What roles do these play in the cell?*

A very interesting activity can be developed by blindfolding a student volunteer and having him hold his nose. A small piece of potato, apple, onion, or other food can be placed in the mouth and chewed. *What is it? How important is smell in identification?*

B. Digestive Processes

Students should be encouraged to interrelate the material previously studied, so that the mechanical and chemical processes of digestion will be highlighted. Each part of the system can be discussed separately and outlined on the chalkboard next to Figure 20-1. The nature of enzymes might be included. The teacher might wish to indicate that enzymes help a reaction occur without being altered by it. The following outline may be reproduced and distributed to students, or it may be placed on the chalkboard:

1) MOUTH

Mechanical—chewing, food pushed by tongue; chemical—saliva moistens food; cooked starches⟶maltose (disaccharide) by the action of the enzyme amylase.

2) PHARYNX, ESOPHAGUS

Food moves to stomach; mucus helps it move; no enzymes.

3) STOMACH

Mechanical—waves of contraction help churn food; chemical—gastric juice; hydrochloric acid, mucus, other secretions (rennin and pepsin) all from the stomach wall, break down proteins to simpler molecules (proteose and peptone); liquified food is now called *chyme*.

4) SMALL INTESTINE

Mechanical—peristalsis, pushes food along; chemical—pancreatic juice from pancreas; amylase digests starch⟶maltose; protease-proteins, proteoses, peptones⟶amino acids; lipase-fats⟶fatty acids and glycerol; from liver-bile-bile salts emulsify fats to prepare for digestion by pancreatic lipase; intestinal juice (*succus entericus*)-protease completes protein digestion to amino acids; sucrase splits sucrose to glucose and fructose; maltase splits maltose to glucose; lactase splits lactose to glucose; absorption-villi absorb 85% of all materials and carry them to the blood.

5) LARGE INTESTINE

Mechanical—moves solid food down through the colon; chemical—no enzymes; mucus present for sliding the materials through; bacteria act on undigested residues; water absorbed.

A number of laboratory activities can be performed. To test for sugars, a plain soda cracker should be broken in half. One half should be crumbled and placed in a test tube with tap water. The other half can be chewed by a volunteer and then placed in a test tube with tap water. Each test tube should be tested with Benedict's solution. Add 2 ml of Benedict's solution to each test tube. *Test tubes should be heated gently, and should never be pointed at anyone while heating. Why?* The presence of carbohydrates will be indicated by a color change from the blue solution to greenish, yellow, orange, or red progressively, depending on the amount of carbohydrate present.

If pepsin extract is available, protein digestion may be studied. A hollow glass tube can be filled with egg white by drawing the fluid into the tube by suction. With one end sealed with a finger, the tube can be transferred to a beaker of boiling water, so that one end is submerged. Contact with the heat will quickly solidify (denature) the egg white. The tube can be removed and allowed to cool. The teacher can use a triangular file to score and break the tube so that a section 1–2 inches long containing solid egg white is made. This can then be placed in a shallow glass dish into which pepsin solution is poured. Digestion of the protein can be seen taking place at the ends of the tube. After several hours, all the protein will have been broken down. The addition of a small amount of hydrochloric acid to the pepsin may speed up the process. *Care should be exercised in the handling of corrosive materials.* Any acid spilled on the skin should be washed off with large quantities of water.

The preceding activity may also be performed using pieces of chopped egg white in test tubes. This provides students with the opportunity to prepare a controlled experiment using water, pepsin, hydrochloric acid, and combinations of these.

The breakdown of fats can be studied by using conditions similar to those which exist in the small intestine. Several test tubes should be prepared by pouring 15cc of water into each. Four drops of oil should be added to each. Olive oil, salad oil, or other similar oils can be used. A few drops of liquid soap should be added to test tube #1, and a few drops of pancreatin solution should be added to test tube #2.

Pancreatin solution can be prepared from pancreatin powder purchased from a commercial supplier, or "intestinal juice" may be prepared by mixing 24cc of NaOH solution (.2M; prepared by mixing 11.2g of NaOH into enough water to make 1 liter); 50cc of K_2HPO_4 solution (.2M; prepared by mixing 40g/liter of solution); 2cc $CaCl_2$ (1% solution; prepared by adding 1g/100cc of water); and 2cc of pancreatin.

The pancreatin solution or intestinal juice should be added to test tube #2 and tube #3. *Are there any controls? What are they?* Each tube should be tested with litmus paper or Hydrion paper. After placing the tubes in a hot water bath for $\frac{1}{2}$ hour, they should be tested again. *Is there evidence of fatty acids?*

3. Uses of the end products of digestion

A. Glucose

The teacher can briefly review the process of the digestion of carbohydrates before introducing the cellular use of glucose in the Kreb's or Citric Acid Cycle. A highly simplified version of this cycle can be presented by using a diagram similar to Figure 20-4. Students should understand that this process is the energy producing process of the cell, and that it is far more complex than indicated.

For advanced groups or as supplementary information, the teacher might choose to include more specific information concerned with energy storage, such as the nature of the ATP (*adenosine triphosphate*) molecule. A simple diagram, such as Figure 20-5, can be used to give a brief explanation of the addition of phosphates to AMP to produce ADP and ATP.

Students should be encouraged to interpret the use of the coiled lines in the diagram to represent bonds of ADP and ATP. The teacher might relate this to the mitochondrion (Lesson 7), the "powerhouse" of the cell. The idea that energy transformation occurs inside the mitochondria and that energy linkage occurs at the surface of mitochondria should be stressed.

The storage of polysaccharides in the form of glycogen in the liver must be mentioned, as well as the storage of glycogen in muscle. *Why is it important that food be available for immediate use?*

B. Amino Acids

A brief review of the function of the endoplasmic reticulum of the cell (Lesson 7) will help students grasp the total concept of protein formation and its relationship to amino acids.

C. Fatty Acids and Glycerol

Student volunteers might mention the structural significance of fats in cell membranes, as well as fat storage and its importance in the body.

The teacher might quickly conclude this lesson by reviewing the stages of digestion and the cellular use for each of the main food groups, and by emphasizing the function of the large intestine or colon in the formation and elimination of solid wastes.

Student reports or class discussion can also be used to introduce pertinent facts about mumps, ulcers, appendicitis, intestinal hernia, hemorrhoids, cirrhosis of the liver, and inflammation of the gall bladder.

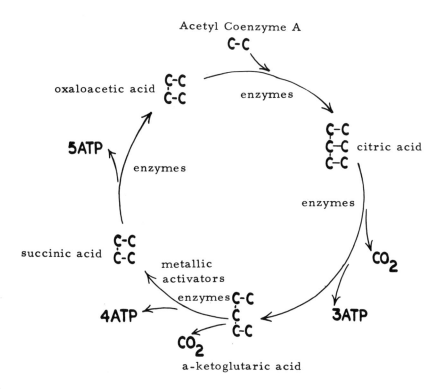

Figure 20-4. Kreb's Cycle.

Figure 20-5. ATP Diagram.

PERTINENT FACTS

- In the human, foods are digested by both mechanical and chemical processes.
- The major portion of chemical digestion occurs in the stomach and upper portion of the small intestine.
- Absorption of digested foods occurs in the small intestine, while water is absorbed in the large intestine.

- Bacteria are essential to the proper functioning of the colon.
- The appendix is a vestigial organ located at the point where large and small intestines join.
- The end products of digestion are used for the production of cell components and for energy production.

POSSIBLE QUIZ

1. Trace the digestion of a cheeseburger, french fries, and a malted milk.
2. What role does peristalsis play in digestion? Does it occur in all parts of the digestive tract?
3. What is the function of saliva, pepsin, trypsin, and HCl in digestion?
4. What is the function of bile? What organ produces it? Where is it stored?
5. How are the end products of digestion used?

READINGS

Beaumont, W., *Experiments and Observations on the Gastric Juice and the Physiology of Digestion* (1833). New York: Dover Publications, Inc., 1959.

Boyd-Orr, J., "The Food Problem," *Scientific American*, August, 1950.

DeCoursey, R. M., *The Human Organism*. New York: McGraw-Hill Book Co., Inc., 1961.

Mayer, S., "Appetite and Obesity," *Scientific American*, November, 1956.

FILMS

"The Human Body: Digestive System." 13 minutes, sound, color, $5.15; bw, $3.65. Coronet Films, Coronet Building, Chicago, Illinois 60601.

"Human Digestion." 10 minutes, sound, bw, $2.15. Soyuztech Film Studios, Contemporary Films, 267 W. 25th Street, New York 10001, N.Y.

"The Multicellular Animal, Part IV: Chemical Machinery." 31 minutes, sound, color, $8.15. McGraw-Hill Book Co., Text-Film Division, 330 W. 42nd Street, New York, N.Y. 10036.

MODELS

1. Plastic assembly models for student use, including the tongue and lower jaw, and the visible man assembly kit, can be obtained from General Biological Supply House, Inc., 8200 South Hoyne Avenue, Chicago, Illinois 60620.
2. Charts and models of the digestive system may also be obtained from General Biological Supply House, Inc.

Lesson 21

THE EXCRETORY SYSTEM AND ITS FUNCTIONS

Lesson time: 45 minutes

Laboratory time: 90 minutes

AIM

To introduce students to the concepts of the functioning of the kidney and the excretion of liquid wastes.

MATERIALS

Sheep or cow kidneys; Urine samples; Benedict's solution; Urinometer or fish tank hydrometer; Distilled water; Filter paper; "Acetest" pills; Sugar and protein test paper; Tablets for testing for bile and blood in urine.

PLANNED LESSON

1. Microscopic kidney structure

In order to make the presentation of information concerning the elimination of liquid wastes more meaningful, the teacher might begin with the structure and function of a single unit of the kidney, the *nephron*. A diagram similar to Figure 21-1 can be developed on the chalkboard as the functioning of the nephron is explained.

Students should propose a possible explanation for the functioning of the nephron according to its structure. The teacher can help them by pointing out the loops of blood vessels surrounding the nephron. *How might liquid wastes move from the blood into the nephron? How does the pressure of the column of fluid in the loop of Henle affect the movement of water molecules out of the loop?* It would also be valuable to point out that much of the fluid in the nephron filters through the tubule walls into the surrounding blood vessels and tissues. Thus, the fluid which remains contains a high concentration of wastes. Students should be able to identify as urine the liquid containing highly concentrated nitrogenous wastes.

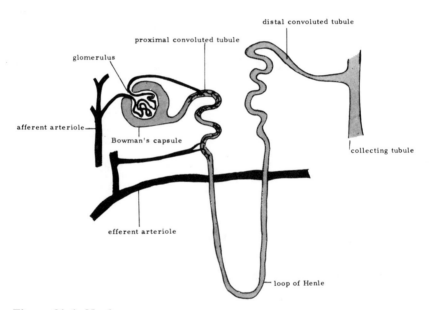

Figure 21-1. Nephron.

2. Gross kidney structure

Using a chart, or a diagram similar to Figure 21-2, the teacher can point out that in longitudinal section, the bean-shaped kidney has two distinct areas, the *cortex* and the *medulla* (made up of parts called pyramids). These can be related to the microscopic structure if students understand that the cortex contains the *glomeruli* of the hundreds of thousands of

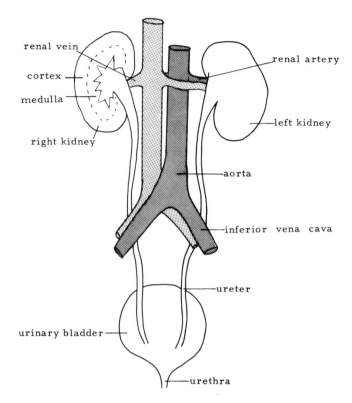

Figure 21-2. Excretory System.

nephrons, while the medulla contains the loops of Henle. The *renal pyramids* can be pointed out and identified as the collection center of all the nephrons of that area. It should be pointed out that a tube called the *ureter* leaves the kidney and connects with the bladder, from which the *urethra* leads externally.

3. Kidney function

The process of urine production might be studied in greater detail. Students should be able to trace the path of blood and the movement of liquids and nitrogenous wastes through the nephron, by using a diagram and reference books. They should pay special attention to the reabsorption of water, and to the distinctions between *threshold* or useful substances and *nonthreshold* substances. The teacher might provide students with a list of substances and ask them to categorize them. The list could include glucose, urea, amino acids, uric acid, sodium, and chloride ions. *At what point in the filtration process are ammonia and urea removed? Where does the reabsorption of glucose, sodium, chloride, and water occur?*

Special student reports might be made on the anti-diuretic hormone (ADH) and its relation to liquid intake and the amount of urine voided; aldosterone; and other influencing factors, as well as other topics of interest. Various diseases and disorders might also be included, such as nephritis, diabetes insipidus, and kidney stones. *What characteristics of kidney function make it possible for man to live in many different environments?*

4. Laboratory activities

The teacher might prepare reference works for the laboratory period including diagrams of microscopic and gross structures of the kidney for labeling; lists of threshold and nonthreshold substances to be identified; and pertinent questions to be answered. If possible, sheep or cow kidneys should be obtained from a local butcher or slaughter house, so that demonstration dissections can be performed by the teacher. Butchers in most food markets will be able to provide such materials *but* it is important to make inquiries well in advance, so that ample time is available.

For advanced or specially interested groups, urine analysis can be done in the laboratory. If the teacher would like to provide his students with an opportunity to test urine which is abnormal, arrangements can usually be made with a local hospital for samples. Normal urine samples can also be obtained from this source. It might be worthwhile for the teacher to provide students with duplicated sets of instructions in advance, so that they will read them over and be familiar with procedures.

A. Specific Gravity

Students can use a urinometer or even a small fish tank hydrometer to determine the specific gravity of the sample they are testing. This can also be performed as a demonstration. They might also determine the reading in distilled water. *Why are the readings different? What does this device measure? How?*

B. Sugar

Benedict's solution might be used here by adding a few drops to 10 cc of urine in a test tube and heating gently. A yellowish-red color indicates the presence of sugar. Commercially prepared tablets or strips can be used in place of Benedict's solution. These are inexpensive and are available in most drugstores. Both tablets and strip come with complete instructions and a color chart. A color change as a result of contact with urine will give an approximate value for the sugar present. *Is sugar a threshold or nonthreshold substance?*

C. Protein

Test paper for protein can also be obtained from a druggist or hospital pharmacy. A color chart is provided. A test solution (sulfo-salicylic acid) may be used. At least 3 cc of liquid and urine should be mixed in a test tube. The liquid will become cloudy if protein is present. *Under normal conditions, can protein molecules pass through cell membranes?*

D. Acetone Bodies

Commercial "Acetest" pills can be used. The pills are packaged with a color chart, and turn purple as a positive result. Student reports might be made to determine the nature of acetone and reasons for its appearance in urine.

E. Occult Blood

Special tablets and filter paper, obtained from a druggist or pharmacy, can be used. The paper should be placed on the table, and 1–2 drops of urine placed on it. The tablet is then placed over this moist spot and a few drops of water added onto the tablet, taking care that some water runs down onto the paper. The color on the filter paper will indicate the presence of blood by turning blue-green. *What reasons might there be for blood in the urine?*

F. Bile

Again, special tablets and paper can be obtained from a local drugstore. The procedure is the same as that for blood testing. *Where do bile salts come from? What might the presence of bile in the urine suggest?*

If local drugstores are not able to supply all needed materials, hospitals or testing laboratories will usually be willing to provide the required supplies.

PERTINENT FACTS

- The nephron is the functional unit of the kidney.
- The kidneys filter nitrogenous wastes out of the body by a process of pressure filtration, selective reabsorption, and tubular secretion.
- Water and threshold substances are reabsorbed by nephron tubules.
- Urine production is controlled by the interrelationship of hormones and fluid intake.
- 99% of the water that filters into the tubules is reabsorbed.
- The kidneys are essential for regulating the internal environment of man.

POSSIBLE QUIZ

1. Sketch the structure of a nephron, and with the aid of arrows, indicate the flow of blood and water through it.
2. Explain the reabsorption process of water.
3. Differentiate between threshold and nonthreshold substances. How does the reabsorption of these occur?
4. List the parts of the excretory system, and briefly explain the role each plays.
5. How does the elimination of waste products in lower animals compare with that of man?

READINGS

Merrill, J.P., "The Transplantation of the Kidney," *Scientific American*, October, 1959.

_____, "The Artificial Kidney," *Scientific American*, July, 1961.

Salisbury, D.F., "Artificial Internal Organs," *Scientific American*, August, 1954.

Smith, H., "The Kidney," *Scientific American*, December, 1953.

FILMS

"The Human Body: Excretory System." 12 minutes, sound, color, $4.15; bw, $2.90. Coronet Films, Coronet Building, Chicago, Illinois 60601.

"The Multicellular Animal, Part IV: Excretion." 27 minutes, sound, color, $8.15. McGraw-Hill Book Co., Text-Film Division, 330 W. 42nd Street, New York, N.Y. 10036.

MODELS

1. Students can construct models of kidney structure using clay or plaster.
2. Charts and models of human kidneys can be obtained from General Biological Supply House, Inc., 8200 South Hoyne Avenue, Chicago, Illinois 60620.

Lesson 22

RESPIRATION IN MAN

Lesson time: 45 minutes
Laboratory time: 90 minutes

AIM

To study the exchange of gases between the lungs, blood and cells.

MATERIALS

Bell jar; Gallon jar; Graduated cylinder; Spirometer; Glass or rubber tubing; Rubber sheeting; Rubber bands; Y-tubing (glass); Rubber stoppers; Balloons; Preserved frogs; Cow or sheep lungs; Scalpels; Alcohol.

PLANNED LESSON

The mechanisms of respiration can be studied by following the process from the inhalation of oxygen into the body, to the ultimate exhalation of the waste product, carbon dioxide.

1. Anatomy of the breathing apparatus

The structure of the respiratory system should include mention of the pharynx, trachea, bronchi, bronchioles, and alveoli of the lungs. A diagram similar to Figure 22-1 can be drawn on the chalkboard, or students may be given copies and asked to label the parts as a research assignment. The location of the diaphragm should also be identified.

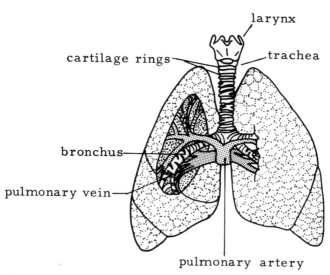

Figure 22–1. Lungs.

2. Physiology of breathing

The teacher might begin by explaining that the quantity of air in the chest cavity is constant, and that the size of the cavity can be changed by the movement of the muscular diaphragm and rib cage. *If the rib cage expands and the diaphragm moves downward, how is the gas pressure on the lungs in the chest cavity altered?* It will be necessary to introduce the concepts of gas pressure to your students. They must clearly understand the idea that *reducing* the size of the chest cavity forces molecules into a more compact arrangement, which in turn means that more molecules hit the surface of the lungs. *Increasing* the space within the cavity can then be explained as readily by students.

The teacher can easily demonstrate this by using a bell jar, a Y-tube, rubber stopper, balloons, and a piece of rubber sheeting. The apparatus should be prepared as illustrated in Figure 22-2. The rubber stopper must be sealed and the balloons must be tied securely. It is advisable to gather a small bit of rubber sheeting at the center and tie a piece of strong string around it, before using rubber bands or string to attach it to the base of the bell jar. This string will allow for pulling gently on the sheet to simulate the downward movement of the diaphragm. The Y-tube may be glass or may be taken from a demonstration stethoscope. Using this set-up, students can simulate the breathing mechanism of the lungs and chest cavity. *What happens to the lungs (balloons) when the diaphragm is pushed up? Why does this happen?*

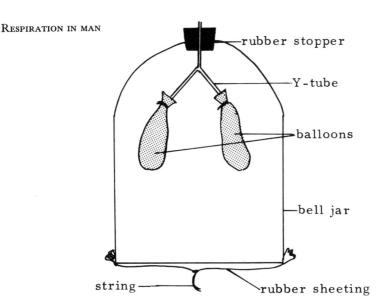

Figure 22-2. Bell Jar Apparatus for Breathing Demonstration.

The teacher should be certain that students understand that the movement of the lungs is a passive process dependent upon the pressure within the chest cavity. Students should also be aware that the movement of the diaphragm is controlled by branches of the phrenic nerve from the medulla. *If someone falls and hits the back of his head, injuring the medulla, what effect might this have on breathing?*

The chemical control of breathing should also be stressed. Students should be aware that the amount of carbon dioxide in the blood stimulates the respiratory center in the medulla, which in turn causes the rate of breathing to either increase or decrease.

3. External respiration

Students must understand that there is a great distinction between "breathing," which is a passive mechanical movement of the lungs, and *respiration*, which is the exchange of gases at the lungs and at the cells. *External respiration* might be explained by student reports or through lecture. The need for a moist lining in the alveoli for gas dissolving, and the close association of the blood capillaries and alveoli should be stressed by these reports. The teacher might here mention that the pigment *hemoglobin* in red blood cells carries the oxygen to the cells (oxyhemoglobin), and carbon dioxide away from them (carboxyhemoglobin). This provides an excellent opportunity to point out the interrelationship of the respiratory and circulatory systems. It might be mentioned that oxyhemoglobin is relatively unstable, while hemoglobin has a greater affinity for carbon monoxide than for oxygen, and the combination of carbon monoxide and hemoglobin is very stable. *How might carbon monoxide poisoning be explained in terms of this bonding?*

4. Internal respiration

In discussing the exchange of gases at the cells, the teacher might mention the Kreb's Cycle (Lesson 20). A brief review of this cycle should be given to highlight the use of oxygen and the production of carbon dioxide.

5. Laboratory activities

A number of interesting and simple activities can be performed by your students.

A. To determine lung capacity, a gallon jar or larger container with a small opening should be filled with water and sealed with a two-hole rubber stopper. Each hole should have a piece of glass tubing inserted into it, making a tight seal. (The inlet tube being near the stopper, the outlet tube *must* extend to the bottom of the jar.) Rubber tubing should then be attached to each glass tube. The end of one tube should be extended so that it rests inside a large graduated cylinder (or beaker on which capacity lines are marked). After swabbing with alcohol or surrounding with sterile material, a student volunteer should blow into the end of the other tube (nose pinched close). Care should be taken to hold the stopper firmly in place. The blowing will force water out of the bottle, displacing the same volume of water as the volume of air which entered. Students may therefore determine the amount of air they were able to expel from their lungs. They can devise tests concerning the effects of hyperventilation and physical exercise on lung capacity.

B. A simple device known as a spirometer can be purchased from most supply houses for use in determination of lung capacity. This will give a reading much more accurate than the method described in the preceding activity. *When performing this activity the nose should be pinched closed. Why?*

It is known that athletic people have a greater vital capacity. *What conditions might be responsible for changing vital capacity?*

C. A number of student-devised tests can be performed to observe factors which alter the respiratory rate. Students should first select a subject and count inhalations per minute. They might then ask the subject to perform some physical activity for a set amount of time, such as running in place for one minute. They should then record the number of respirations for one minute immediately after exercise. After an adequate rest, the subject might be asked to hyperventilate and repeat the activity. *Does this affect respiratory rate?* A number of other tests can be devised from these. A comparison of rates in athletes and nonathletes; smokers and nonsmokers, should prove interesting.

D. The lungs and trachea can be removed from a preserved frog, or cow or sheep lungs may be obtained from a meat packer or butcher. A proper sized glass or rubber tube can be attached to the trachea, so that a student can blow into the lungs causing expansion of the tissue. The lungs and trachea may later be cut open to observe internal structure.

A natural conclusion to this lesson would be to have students discuss and point out the interrelationships between the respiratory and circulatory systems.

PERTINENT FACTS

- The filling and emptying of the lungs is a passive process dependent upon the gas pressure in the chest cavity as it is controlled by the diaphragm and the muscles of the chest.
- Oxygen is picked up by the blood in the alveoli of the lungs.
- The exchange of gases at the lungs is called external respiration.
- The combination of the respiratory pigment hemoglobin and oxygen forms a relatively unstable compound known as oxyhemoglobin.
- An excess of CO_2 in the blood stimulates the respiratory center in the medulla to speed up the rate of breathing, while a drop in the level of CO_2 causes the medulla to decrease the rate of breathing.
- Oxygen is used in the production of energy in the Kreb's Cycle, at which time carbon dioxide is given off as a waste product.

POSSIBLE QUIZ

1. Explain the process of breathing and indicate the factors involved.
2. Differentiate between internal and external respiration and explain their interrelationships.
3. How is the circulatory system related to the respiratory system? What is the role of hemoglobin in the transport of gases?
4. How does the activity of the cell affect the rate of respiration?
5. How would high altitudes affect respiratory rate and gas transport?

READINGS

Chaffee, E.E., and E.M. Greisheimer, *Basic Physiology and Anatomy.* Philadelphia: J.B. Lippincott Co., 1964.

Clements, J.A., "Surface Tension in the Lungs," *Scientific American,* December, 1962.

Comoe, J.H., "The Lung," *Scientific American,* February, 1966.

Fox, H.M., "Blood Pigments," *Scientific American*, March, 1950.

Krogh, A., *Comparative Physiology of Respiratory Mechanisms*. Philadelphia: University of Pennsylvania Press, 1959.

Perutz, M.F., "The Hemoglobin Molecule," *Scientific American*, November, 1964.

FILMS

"The Human Body: Respiratory System." 12 minutes, sound, color, $4.15; bw, $2.90. Coronet Films, Coronet Building, Chicago, Illinois 60601.

"Mechanisms of Breathing." 11 minutes, sound, bw, $2.15. Encyclopedia Britannica Films, Inc., 1150 Wilmette Avenue, Wilmette, Illinois 60091.

"The Multicellular Animal, Part IV: Respiration." 27 minutes, sound, color, $8.15. McGraw-Hill Book Co., Text-Film Division, 330 W. 42nd Street, New York, N.Y. 10036.

MODELS

1. Display mounts, models, and charts of the respiratory system can be obtained from General Biological Supply House, Inc., 8200 South Hoyne Avenue, Chicago, Illinois 60620.

Lesson 23

BLOOD AND CIRCULATION

Lesson time: 45–90 minutes

Laboratory time: 90 minutes

AIM

To study the nature of blood and the anatomy and physiology of the circulatory system.

MATERIALS

Microscopes; Glass slides; Cow or sheep hearts; Blood sample; Blood typing sera; Wright's stain; Alcohol; Disposable lancets; Sterile cotton; Tallqvist scale filter paper; Toothpicks; Scalpels; Film.

PLANNED LESSON

1. Blood

Before discussing the structure and functioning of the heart and blood vessels, students should understand the fundamentals of blood composition and its nature. The teacher might begin by differentiating between the liquid portion, or *plasma* of blood, and the cells found in the plasma.

A. PLASMA

A small tube or other container of blood might be obtained from a local hospital or laboratory. By allowing this to remain undisturbed for a day or two, the teacher will be able to illustrate the settled cells and the straw-colored plasma. It would be valuable to have students do individual research to identify the components of the plasma and their importance. They should include water as the chief component; plasma proteins such as albumin and fibrinogen; blood glucose; electrolytes such as sodium, potassium, chlorides, phosphates, and bicarbonates; and other substances, including the nitrogenous wastes uric acid, urea, and creatin; various hormones and enzymes; and the respiratory gases. The volume of blood might also be mentioned.

B. Cells

The various types of blood cells can be presented separately.

1) The red blood cells or *erythrocytes* can be observed by having students make a blood smear slide and examining it under the microscope. Proper technique for taking blood samples must be used in all activities, and must be carefully followed by students. The teacher should always check with his department chairman so that standard procedures are followed. It is always necessary to have students bring a note signed by parents or guardian granting permission to participate.

The materials which will be needed include sterile, individually packaged disposable blood lancets, alcohol, sterile cotton, and clean glass slides. (Reusable plungers of any kind must *never* be used since diseases such as hepatitis can be transmitted in this way.) Care should be taken that each time a lancet is used, it is *immediately disposed* of.

The procedure should be explained before beginning, and it is suggested that in most classes the teacher do the actual skin puncture. Students should be instructed to wash their hands thoroughly, dry them, and swab the third finger of one hand with a piece of cotton saturated with alcohol. That arm should then be held in the extended position downward and gently shaken, while the alcohol evaporates.

Immediately before puncturing, the finger should be gently squeezed with the thumb and index finger of the other hand, in a direction from its base to the tip. The movement should stop before contacting the swabbed area. It will be seen that the tip becomes dark red as this "milking" motion causes blood to collect. The teacher should grasp the student's hand firmly, so that he is holding the third finger securely. In order to prevent possible jerking of the arm, it is advisable to have the student stand to the rear left of a right-handed teacher, or rear right for a left-handed teacher. In this manner the teacher can hold the student's arm to the teacher's side by pressing the student's arm with the teacher's elbow.

The arm will then be relatively immobile. (The hand should never be placed on a surface, such as a table top, when being punctured. This may result in too deep a puncture and possible bone nicking.)

Holding the finger in the teacher's hand so that pressure on the tip keeps it engorged with blood, the teacher can use a quick movement to puncture the tip with the lancet, approximately $\frac{1}{4}$ inch behind the nail. Make certain that the lancet has been unwrapped in such a manner that the tip has not touched or been touched by any object. The lancets should be placed in a receptacle as each is used to prevent any student from attempting to take one for later use. The teacher should later dispose of these safely.

By turning the hand so that the nails are upward, maintaining squeezing pressure, and holding the finger above a slide, a drop of blood can be transferred to the slide. The finger should not touch the slide. The student should then be given a cotton swab saturated with alcohol to place over the puncture. Using the thumb, the student can hold the cotton in place. The pressure needed for this will also aid in stopping bleeding, which will be slight in any case. Not all students should have their fingers punctured for this test, since a number of other tests will also require samples.

To study red and white blood cells, students should spread the drop of blood across the slide by using the narrow edge of a second slide to pull the blood along the slide. Figure 23-1 illustrates this procedure. They should be able to sketch the structure of these cells, and be encouraged to observe carefully. *What organelle seems to be missing from these cells?*

The teacher may wish to go into a discussion of the theories which suggest that since red blood cells contain no nuclei (in man), they are really not cells. Advanced or specially interested students might prepare a report on this. If available, a film such as "Heart and Circulation" can be used to show the movement of red blood cells through the capillaries, as well as their *rouleaux arrangement*.

A special study should be made of hemoglobin. The union of oxygen and carbon dioxide with hemoglobin in the process of respiration may previously have been studied in respiration, and should be reviewed. The teacher might link the previous study of protein synthesis at the endoplasmic reticulum, by indicating that this pigment is a complex protein produced within the red blood cell. The incorporation of iron within the hemoglobin structure should also be mentioned. *What is iron deficiency anemia?* It should also be noted that both red and white blood cells are produced by the bone marrow primarily in the long bones and sternum. *What diagnostic value might a sternal puncture have?*

When a drop of blood is taken for the preparation of a smear, a second

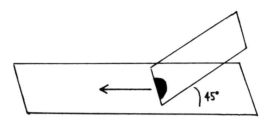

Figure 23-1. Blood Smear Slide.

drop might be placed on a sheet of filter paper of the Tallqvist scale. Again, the drop should touch the paper but the finger should not. The book of paper contains a color chart which can be used as an approximation of the amount of hemoglobin in the blood. The drop should be allowed to dry just to the point at which it is no longer shiny and then be compared with the chart. *Why is this not an accurate test? What is anemia? How are the amount of hemoglobin and the number of red blood cells related?*

Blood typing can also be done in the laboratory at this time. The teacher should explain that the major blood groups react differently when tested, and might place a chart similar to Figure 23-2 on the chalkboard, indicating these reactions. The plus signs on this chart indicate that clumping or agglutination occurs.

Anti A and anti B sera can be purchased from a pharmacy or may be obtained, outdated, from a hospital or laboratory. A line, drawn with a wax pencil, can be used to divide a glass slide into two halves. A drop of anti A sera should be added to the left side and a drop of anti B to the right side. Two small drops of blood can be placed in each drop of sera. Using a clean end of a toothpick for each drop, they should be stirred so that blood and serum mix. The agglutination will be clearly visible. Students should be warned that this testing is only a demonstration, the results of which should not be used for any diagnostic purposes.

The teacher may go into the mechanisms behind this agglutination process by relating the agglutinogens of red blood cells to the agglutinins of plasma. *What blood group would be the universal donor? The universal recipient? What will be the effects of transfusing the wrong type blood?*

2) *Leukocytes. What does* leukocytes *mean?* Another member of the laboratory group can be used to obtain a drop of blood. The blood should be spread across the slide as in Figure 23-1, dried, and Wright's stain added for three minutes. Add several drops of distilled water and blow on this mixture until a metallic film forms. Five minutes later, wash off gently with tap water, and observe under the microscope *Do all white blood cells look alike? How can they be identified?* Students should have no difficulty

observing neutrophils and may also see monocytes and basophils. They should use reference materials to help in identification.

The origin of white blood cells in the spleen, lymph nodes, and tonsils, as well as the bone marrow, will help students to understand the abnormalities of these in blood-related diseases (as in the enlargement of the spleen in leukemia patients).

The movement of white blood cells toward an invading substance (*chemotaxis*) can be used to introduce the phagocytic nature of these cells. The process of engulfing materials by phagocytosis may have been previously covered (Lesson 6) and can be reviewed here. Students should be able to explain that the destruction (digestion) of these substances would probably occur as the result of the action of enzymes. Student reports concerning the various types of white blood cells might be asked for, although they should be warned that little is known about some of them.

3) *Platelets (thrombocytes)*. The lack of information concerning the origin and life cycle of these cells can be used by the teacher to remind students of the many gaps which exist in scientific knowledge. They should be aware that researchers do not know "everything about everything," and perhaps more important, that theories are modified and changed by new evidence.

BLOOD TYPE	ANTI-A	ANTI-B
A	+	−
B	−	+
AB	+	+
O	−	−

Figure 23-2. Blood Agglutination Chart.

The function of platelets in blood clotting should be mentioned. A chart similar to Figure 23-3 might best be used to illustrate the stages involved in the clotting process. The role of calcium within this process can be used to emphasize the use of minerals within the body. Students should understand the need to prevent clotting *within* the blood vessels. *What is the importance of inhibition by substances such as heparin?* The probable lack of antihemophilic factor in the bodies of hemophiliacs can be traced through the cycle, so that the resultant hemorrhaging in such people will be understood.

2. The heart

The "pump" of the circulatory system can be studied, and the size, location, and general appearance given. *What is the function of the pericardium?* Using a chart or a chalkboard illustration similar to Figure 23-4, the teacher can point out the atria, ventricles, valves, and major vessels which enter and leave the heart. The path of the blood flow can be indicated with arrows. The functioning of the valves as it is related to their shape should be stressed. *Why does the contraction of the ventricles cause the bicuspid and tricuspid to close, while the semilunar valves open?* A sheep or cow heart can be used for dissection to demonstrate the structure of the heart. Adequate advance notice must be given to the butcher or slaughterhouse (abattoir).

The pumping of the heart, or heartbeat, can be considered in some greater detail. Students are no doubt already familiar with the "lub-dub" of the heartbeat. They should be able to relate these sounds to the contraction and relaxation of the heart. Using stethoscopes, students will be able to listen to their hearts and to those of their classmates (To really hear sounds, a very quiet room must be used and the stethoscope should be

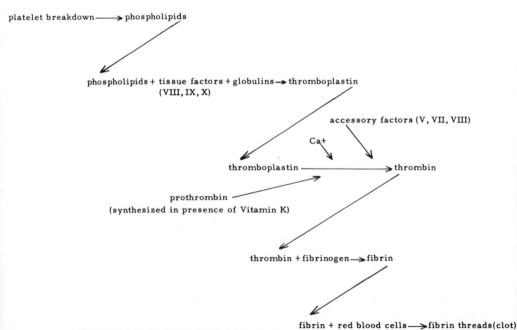

Figure 23-3. Blood Clotting Diagram.

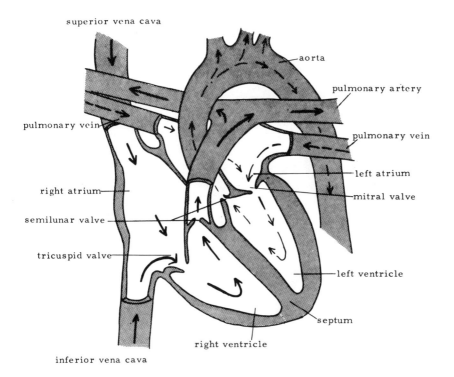

Figure 23-4. Heart.

placed next to the skin.) Using reference books, they can determine the nature of the *systole* and *diastole*. In discussing the built-in conduction system of the heart, the Bundle of His, atrio-ventricular node, sino-atrial node, and Purkinje fibers should be mentioned. The origin of the impulse and its transmission can be shown in outline form:

SA node → atria → triggers AV node → Bundle of His → Purkinje fibers. (A diagram such as Figure 23-5 might also be used).

The triggering of the AV node by the atrial conduction of the impulse can help students to understand the coordination of total heart contraction. *What might happen if impulses were initiated independently in the atria and ventricles?* The teacher should also mention the control of the heartbeat by the parasympathetic and sympathetic branches of the vagus nerve. *What happens when faster impulses are sent through the parasympathetic branch?*

The four-chambered heart of man can be used to illustrate advances in the evolutionary development of this organ. Interested students can prepare reports concerned with electrocardiograph techniques, blood pressure, heart ailments, heart-lung machines, and heart transplants.

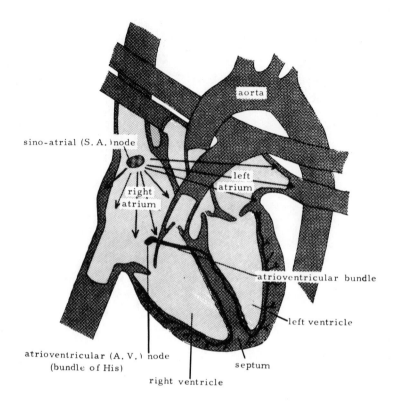

Figure 23–5. Heart Innervation.

3. Blood vessels

Moving from the pump to the "pipelines," the arteries, veins, arterioles, venules, and capillaries will complete the study of this system. Some major vessels should be mentioned, such as the aorta, vena cavae, and pulmonary vessels. Students should clearly understand that arteries carry blood *away* from the heart, and veins carry it to the heart. *Since blood is oxygenated just before it leaves the heart, what statements can be made about the oxygen content of the blood in the veins? In the arteries?*

Here, the teacher should be certain to emphasize the exception to this. That is, the pulmonary artery carries blood away from the heart but this blood is low in oxygen, while the pulmonary veins, unlike other veins, carry blood high in oxygen. *Considering the energy expended by heart tissue, why are the cardiac vessels which supply heart muscle among the first to branch from the aorta?*

The branching of arteries to arterioles and finally to capillaries, and the enlargement of these to venules and veins, can be used as a basis for the study of general circulation. Bringing in the absorption of food in the digestive system, the transfer of gases at the lungs and the passage of materials into and out of the cell, the need for very thin-walled vessels (capillaries) can be stressed. The great elasticity of vessels should also be mentioned. *Why does a cut artery spurt blood, while a vein bleeds steadily? Why do doctors usually tie a rubber tube around the upper arm before taking blood from the vein?*

4. Lymph

The teacher should include a discussion of this fluid, which is the result of intracellular metabolism, or has exuded through the capillaries. As this blood plasma does not all move back into the capillaries, its existence in the intercellular spaces should be mentioned. The lymph vessels cannot be seen easily in the laboratory, but the class may contain some student who has had the characteristic "red line" moving up the arm as a result of infection of a finger. The swelling of lymph nodes, like "swollen glands" of the neck, might also be considered. The blockage of lymph vessels by filarial worms and the resultant *elephantiasis* might prove an interesting study for a student volunteer.

PERTINENT FACTS

- Blood is composed of plasma, red cells, white cells, and platelets.
- Red blood cells are chiefly involved with carrying respiratory gases, while white blood cells are responsible for the destruction of foreign substances.
- Blood platelets, and substances in the liver and plasma, are involved in blood clotting.
- The heart has its own mechanism for the initiation of contractions, as well as being affected by branches of the sympathetic and parasympathetic nervous system.
- The systemic circulation carries blood to all parts of the body except the lungs, while the pulmonary circulation involves the passage of blood toward and away from the lungs.
- Lymph is blood plasma which has filtered out of the capillaries into the intracellular spaces. Lymph contains 50% less protein than plasma.

POSSIBLE QUIZ

1. Trace the path of blood through the heart, and at each step, indicate whether it is oxygenated or deoxygenated.
2. Trace the path of a drop of blood from the right big toe to the left ear lobe.

3. How are the rhythmic contractions of the heart initiated? How are they controlled?
4. What are the functions of blood plasma, erythrocytes, leukocytes, and platelets?
5. Outline the steps involved in blood clotting. What is the importance of the inhibitory effect of heparin?

READINGS

Bing, R.J., "Heart Metabolism," *Scientific American*, February, 1957.

Debakey, M.E., and L. Engel, "Blood Vessel Surgery," *Scientific American*, April, 1961.

Ebert, J.D., "The First Heartbeats," *Scientific American*, March, 1959.

Kilgour, F.G., "William Harvey," *Scientific American*, June, 1952.

Kolf, W.J., "An Artificial Heart Inside the Body," *Scientific American*, November, 1965.

Myerson, H.S., "The Lymphatic System," *Scientific American*, June 1963.

Ponder, E., "The Red Blood Cell," *Scientific American*, January, 1957.

Slaughter, F.G., "Heart Surgery," *Scientific American*, January, 1950.

Surgenor, D.M., "Blood," *Scientific American*, February, 1954.

FILMS

"Heart Disease: Its Major Causes." 11 minutes, sound, bw, $2.15. Encyclopedia Britannica Films, Inc., 1150 Wilmette Avenue, Wilmette, Illinois 60091.

"Heart: How It Works." 13 minutes, sound, bw, $2.65. McGraw-Hill Book Co., Text-Film Division, 330 W. 42nd Street, New York, N.Y. 10036.

"Heart, Lungs, and Circulation." 11 minutes, sound, color, $3.40; bw, $2.15. Coronet Films, Coronet Building, Chicago, Illinois 60601.

"The Multicellular Animal, Part IV: Circulation." 27 minutes, sound, color, $8.15. McGraw-Hill Book Co., Text-Film Division, 330 W. 42nd Street, New York, N.Y. 10036.

MODELS

1. Charts and models of the heart and circulatory system may be obtained from Ward's Natural Science Establishment, Inc., P.O. Box 1712, Rochester, New York 14603; and from General Biological Supply House, Inc., 8200 South Hoyne Avenue, Chicago, Illinois 60620.
2. Plastic assembly models and kits for student use, including heart model kit, and *functional* heart model kit may be obtained from General Biological Supply House, Inc.

Lesson 24

THE ENDOCRINE SYSTEM

Lesson time: 45 minutes

Laboratory time: 90 minutes

AIM

To develop an understanding of the production of hormones and their control of body functions.

MATERIALS

Tadpoles; Pond or aquarium water; Tincture of iodine; Bits of boiled lettuce; Films.

PLANNED LESSON

1. The nature of hormones and their role in homeostasis

The study of the endocrine system might begin with a discussion of the chemical messengers produced by the ductless or *endocrine* glands. The meaning of the word hormone, *I arouse*, can be used to explain the functions these perform. Students can be asked to use reference books to establish a working definition for the term. The class definition should include the idea that these are organic substances secreted by endocrine glands, and are carried by the blood to specific or general sites in the body, where they exert a regulatory influence. They should understand the concept that some hormones chemically regulate many activities; others regulate body growth; others control sexual development; others act as catalysts to help other processes; while still others control the cellular use of glucose, fats, carbohydrates, salts, and water in the maintenance of *homeostasis*.

Interested students may wish to report on the various methods of study by which the role of individual hormones can be identified. They might include reports on the microscopic appearance of normal gland tissue; effects of the removal of a particular gland in laboratory animals; the results of experimentation during which hormones are injected into laboratory animals; and the analytical methods used in laboratories to study hormones, blood, and glands. Reports on specific researchers, such as Banting and Best, and Bayliss and Starling, should prove very interesting.

2. Glands and their functions

In order to introduce the study of the endocrine glands and their functions, the teacher might prepare an outline chart, duplicate it and distribute it to students to be completed with the aid of reference materials. The column headings should include: gland, location, size and shape, hormones produced, regulatory action of hormones, abnormalities, and control by the pituitary. The glands to be included might be pituitary, ovaries, testes, thyroid, parathyroids, adrenals, the islets of Langerhans of the pancreas, and the hormone producing glands of the digestive tract. The teacher can use the completed charts as the basis for a discussion of *feedback*. (One of the finest feedback mechanisms known involves the various hormones of digestion.) The students should have already become aware of the interrelationships among the hormones studied. *How does an increase in thyroid-stimulating hormone affect the homeostasis of the body?* The film "Endocrine Glands" might be shown to clarify and reinforce the material covered.

3. Abnormalities

Some special attention should be given to this aspect of the lesson. Some specific conditions which can be studied are acromegaly, diabetes insipidus, cretinism, goiter, diabetes mellitus, and Addison's disease.

4. Laboratory activities

An interesting laboratory activity involves the hormonal effects on metamorphosis. Add 18-20 drops of tincture of iodine to a liter of pond water containing three large tadpoles whose hind-limb buds have just become visible. Add bits of boiled lettuce, taking care to change the water and the tincture of iodine every other day. In a second liter of pond water, keep three other tadpoles whose hind-limb buds also have just become visible. Feed with bits of boiled lettuce, but *do not* add iodine. *What is the purpose of this second set of tadpoles? How long does metamorphosis take to occur in the iodine-fed tadpoles? In the non-iodine-fed tadpoles? How can you account for this?*

Students can carry this experiment further, by varying the size of the tadpoles and the amount of iodine used. In this manner, they can try to determine the amount of iodine needed for optimum growth and development, and also the stage of development at which iodine no longer exerts an effect. *What is the relationship between iodine and thyroxine? Could you determine what concentration of iodine is best suited to development?*

PERTINENT FACTS

- The endocrine, circulatory, excretory, and nervous systems serve to control and integrate the various activities of the body.
- In general, the endocrine glands are ductless, and liberate their hormonal secretions directly into the bloodstream.
- Hormones are chemical messengers which are carried by the blood to various parts of the body, where they exert their specific effects.
- The anterior lobe of the pituitary gland secretes growth hormone, gonadotrophic hormone, ACTH and TSH.
- The posterior lobe of the pituitary secretes the hormones vasopressin, which controls diabetes insipidus, and oxytocin, which produces contractions of the walls of the uterus.
- Diabetes insipidus is a condition in which the kidneys fail to concentrate the urine as a result of decreased production of vasopressin.
- In addition to the sperm producing functions, the testes also secrete testosterone and other *androgenic* hormones serving to stimulate development of secondary sexual characteristics in the male.
- Estrogen, an ovarian hormone, is the counterpart to testosterone in the female.
- The ovaries also secrete the hormone progesterone, essential for complete development of the reproductive structures and the establishment of the menstrual cycle.
- The thyroid hormones, of which thyroxine is the most abundant, stimulate the rate of oxidative metabolism of the various cells and tissues of the body.
- The secretion of the thyroid hormones is controlled by the thyroid-stimulating hormone (TSH), secreted by the anterior pituitary gland.
- In a newborn child an underactive thyroid results in cretinism, while

in adults the commonest forms of *hypothyroidism* are myxedema and endemic goiter.
- The commonest forms of *hyperthyrodism* are exophthalmic goiter and Grave's Disease.
- The function of the parathyroids is to maintain the proper balance of calcium and phosphorus in the blood.
- The parathyroid glands secrete the hormone parathormone in response to the calcium level of the blood, and function to maintain the normal excitability of the neuromuscular system.
- Adrenalin or epinephrine is produced by the adrenal medulla, while the adrenal cortex produces the hormones cortisone and aldosterone.
- In man, Addison's disease is the commonest form of adrenal cortex insufficiency, and is associated with an increased excretion of sodium salts by the kidneys.
- The development and function of the adrenal cortex is under the control of a hormone produced by the anterior pituitary, known as ACTH.
- The hormones insulin and glucagon, secreted by the pancreas, play a role in regulating the carbohydrate metabolism of the body.
- Insulin decreases the concentration of sugar in the blood, while glucagon has the opposite effect.
- A lack of sufficient insulin production by the pancreas results in diabetes mellitus.
- Gastrone, enterogastrone, cholecystokinin, and secretin are some hormones involved in digestion.

POSSIBLE QUIZ

1. For each of the following conditions list the gland or glands in question, the hormones involved, and whether the gland is hypo- or hyper-active:
   a. Diabetes insipidus
   b. Diabetes mellitus
   c. Cretinism
   d. Addison's Disease
   e. Grave's Disease.
2. What possible benefit might be gained by the injection of oxytocin into a pregnant woman whose labor is overdue?
3. Select any of the hormones studied, and use it to illustrate the mechanisms of feedback.
4. Discuss the various ways in which hormones exert their effects on the body. How have researchers been able to identify these functions?
5. Describe the nature of hormones. Why are they sometimes referred to as chemical messengers?

READINGS

Gray, G.W., "Cortisone and ACTH," *Scientific American*, March, 1950.
Levey, R.H., "The Thymus Hormone," *Scientific American*, July, 1964.
Li, C.H., "The ACTH Molecule," *Scientific American*, July, 1963.
————, "The Pituitary," *Scientific American*, October, 1950.
Rasmussen, N., "The parathyroid Hormone," *Scientific American*, April, 1961.
Thompson, E.O.P., "The Insulin Molecule, "*Scientific American*, May, 1965.
Wilkins, L., "The Thyroid Gland," *Scientific American*, March, 1960.
Wurtman, R.J., and J. Axelrod, "The Pineal Gland," *Scientific American*, July, 1965.
Zuckerman, S., "Hormones," *Scientific American*, March, 1957.

FILMS

"Endocrine Glands." 11 minutes, sound, bw, $2.15. Encyclopedia Britannical Films, Inc., 1150 Wilmette Avenue, Wilmette, Illinois 60091.

"The Multicellular Animal, Part IV: Hormones." 27 minutes, sound, color, $8.15. McGraw-Hill Book Co., Text-Film Division, 330 W. 42nd Street, New York, N.Y. 10036.

MODELS

1. Microscope slides and color transparencies of the endocrine glands can be obtained from Carolina Biological Supply Co., Burlington, North Carolina 27215.
2. Charts and models of the endocrine glands may be obtained from Ward's Natural Science Establishment, Inc., P.O. Box 1712, Rochester, New York 14603.

Lesson 25

NERVOUS AND SENSORY MECHANISMS

Lesson time: 135 minutes

Laboratory time: 90 minutes

AIM

To develop an understanding of the structure and function of the nervous system as well as the special senses.

MATERIALS

Human skull; Eye and ear models; Preserved cow or sheep eyes; Rubber ball; Tuning fork; Sheet of white paper; Sheet of black paper; Blunt pins; Sterile cotton.

PLANNED LESSON

The lesson may be introduced by considering the major subdivisions of the nervous system. A brief explanation of terms might be given at this time. The *central nervous system* should be presented first. The brain and spinal cord should be suggested by your students from the name alone. *Where are each of these located?* The *peripheral* system can then be identified by students, considering the meaning of the term "periphera."

The terms *somatic, visceral, afferent,* and *efferent* can be introduced at this point by having students use reference materials to identify them, or by developing their meanings through discussion. Each division might then be considered separately, with attention to both microscopic and macroscopic arrangement.

1. Central nervous system: the spinal cord

The teacher might use a disarticulated skull to identify the *foramen magnum*, or big hole, at the base of the skull. *What function does this serve?* Cross-sectional study of the cord can be included here if the teacher feels it of value. The *reflex arc* should be mentioned as well. The teacher might use a demonstration to illustrate reflexes at work. If a rubber ball is tossed at a student nearby, he will attempt to shield himself or catch the ball (reflex). Discussing this, as well as many other similar situations (hand on a hot stove, eye blinking and others), students might be asked: *Did the subject consciously think about the situation and then react? Did he react and then think about the situation?* This should help bring out the concept that we do many things without conscious thought. The teacher can now introduce the actual mechanism involved. It will probably be sufficient to indicate that the path includes a *receptor*, an afferent nerve carrying the impulse to a center, as in the spinal cord, and an efferent nerve carrying an impulse to an effector. Students can then be asked to explain, in these terms, what happens when we touch something very hot.

Students should be able to develop the second function of the cord by analyzing its location. Thus, they will be able to understand that it conducts impulses to and from the brain. In advanced groups, the teacher might want to include some study of the ascending and descending fiber tracts, although it is not necessary for average groups.

The *meninges* should also be mentioned. Students can indicate that the cord is protected by the spinal column or vertebrae. The teacher can simply indicate that the *dura mater* (hard mother) is the outer covering, the *arachnoid* (spider web) layer is next, and the *pia mater* (gentle mother) is innermost, and in contact with the cord. *What would a subdural hemorrhage be? What is meningitis?*

2. Central nervous system: the brain

Because the brain is highly complex, it would not be valuable for the teacher to go into great detail concerning its structure or function. A sketch similar to Figure 25-1 can be used to indicate parts of the brain, as well as to outline the entire central nervous system. If the teacher

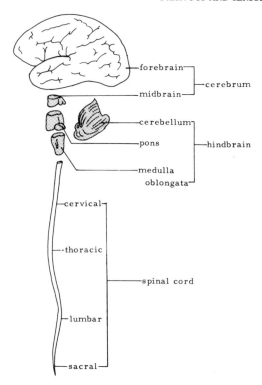

Figure 25-1. Central Nervous System.

wishes, he can obtain cow or sheep brains from a local butcher or market. These, together with brain models, can be used to demonstrate the various brain areas.

A. Brain Areas

A study of the functions of the major areas of the brain can be introduced. The teacher might begin with the *medulla oblongata*, and consider the role of its "vital centers" of heart, blood pressure, respiration, and control of swallowing. *What might occur if someone falls and hits the back of his head so that the medulla is stunned? Considering that the phrenic nerve emanates from the medulla, what is the respiratory control involved?*

B. Cerebellum

The coordinating role of the cerebellum may be stressed by considering the control of muscle activity in voluntary movements. Its functioning in relation to proprioception can also be included. Some interesting activities to test proprioception and corrections made by the cerebellum are included in the Laboratory Activities of this lesson.

C. Diencephalon

You may wish to mention the structures found beneath the cerebrum, the *hypothalamus* and *thalamus*. It will probably be sufficient to indicate that the hypothalamus regulates visceral activities, and the thalamus is an integrating center of impulses. Individual students might present oral reports on these as a special activity.

D. Cerebrum

Students can use commercially prepared charts or other reference materials to study the anatomy of this largest part of the brain. In the laboratory, sheep or cow brains can be observed and dissected, if available. *What are the cerebral hemispheres? What is the corpus callosum?*

A sketch similar to Figure 25-2 can be duplicated and distributed to your students so that they can indicate the various areas of the lobes and fissures. You may also wish to supply your students with a list of terms to be used, such as, temporal, frontal, parietal, and occipital lobes; fissures of Rolando and Sylvius; and the various senses and the area which controls them. For some groups, it will be best to simply ask that they label the sketch as completely as possible. *What is the cerebral cortex? What is its function? What is sleep?*

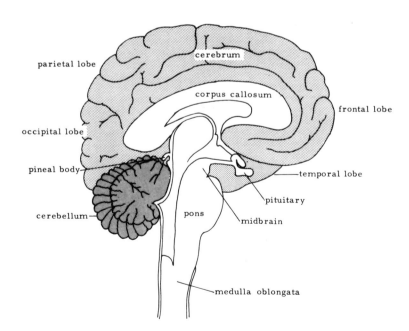

Figure 25-2. Brain.

E. Meninges

As students have previously been introduced to the meninges as the covering of the brain, it will be necessary simply to explain their continuous nature to include the spinal cord. A brief discussion of cerebrospinal fluid should also be included. *What are the functions of this fluid? What is a spinal tap? Why is it necessary to keep the patient prone to prevent a headache after spinal anesthesia?*

Individual students might wish to present special reports to the class dealing with such subjects as meningitis, epilepsy, multiple sclerosis, polio, neuralgia, delirium, encephalograms, or the results of injury to various parts of the nervous system.

3. Peripheral nervous system

The study of the nervous system can include a close look at the cells which compose it and how these cells function. Diagrams similar to Figure 25-3, of the microscopic structure of "typical" neurons, can be used to identify the cell body, Nissl granules, neurofibrils, dendrites, and axons. Differentiation should be made between myelinated and *non*myelinated neurons.

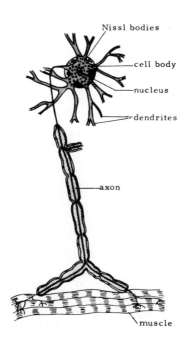

Figure 25-3. Motor Neuron.

Students can use references to identify bipolar and multipolar neurons, as well as afferent, efferent, and internuncial types. *How are neurons classified? In what direction does a nerve impulse travel through an axon? A dendrite?*

Students should understand that the peripheral nervous system is composed of a connecting and branching network of fibers and nerve cells, extending to all parts of the body. *What is the function of the cranial and spinal nerves of the peripheral nervous system?*

4. Characteristics of nervous tissue

A. EXCITABILITY AND CONDUCTIVITY

The teacher should include a study of excitability and conductivity. *What is the all-or-none response? What is the threshold stimulus?* Some information concerning the chemical nature of impulse conduction might be included here. Comparing the nerve to an electrical system, the teacher can introduce the idea of a need for an electrical potential on the axon membrane. Here again, the role of salt ions within the cell can be stressed.

The idea that a "sodium pump" mechanism moves sodium out of the neuron, thus creating a difference in electrical charge on the membrane, might be included. In most cases it will be sufficient to simply identify the movement of these ions back into the neuron, as the factor which causes the movement of an electrical (nerve) impulse along the membrane. Figure 25-4 should prove helpful in illustrating this concept. *If the membrane of a resting nerve fiber has a positive charge on the outside and a negative charge on the inside, what would be the effect of stimulation?*

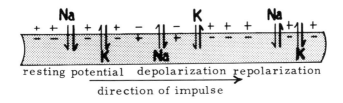

Figure 25-4. Transfer of Impulse Along a Nerve.

B. THE SYNAPSE (NEUROMYAL JUNCTION)

Students should be familiar with the area at which the *dendrites* of one neuron come in close contact with the *axon* of another. It should be stressed that there is no actual contact between them, but only a contiguity or closeness. The role of acetylcholine in impulse conduction across this synaptic gap and the antagonistic role of cholinesterase should be included in discussion. *What might happen if cholinesterase was not present at the synapse?*

5. Autonomic nervous system

Although a somewhat artificial division, the visceral efferent fibers of the peripheral nervous system might be considered separately as they are usually regarded so by scientists. Students should understand that in fact, all the functions of the nervous system are closely interrelated. Students will need to know that the cell bodies of nerves of the autonomic nervous system are within the brain or spinal cord, and originate in only three areas: the cranial, thoracolumbar, and sacral regions. It must also be clear that the fibers from these areas actually combine to form two divisions, the *craniosacral* and the *thoracolumbar*. A diagram similar to Figure 25-5 will prove useful in presenting this information.

Reference work or class discussion should bring out the fact that many visceral effectors have a double nerve supply, one from each of these divisions. The teacher should stress the antagonistic functioning of these systems and the balance maintained by the braking action of one and the accelerating action of the other. Examples of the functioning of these can include the stimulating effect of thoracolumbar impulses on the heart and the inhibiting effect of the craniosacral; the dilation of the iris by the thoracolumbar and contraction of the iris by the craniosacral; the pilomotor muscle control (goose bumps) of the thoracolumbar; and the increase in the breakdown of glycogen leading to more available glucose as a result of thoracolumbar stimulation. Using these and other examples if necessary, students should be able to indicate the division which is called the *sympathetic* system and the one known as the *parasympathetic* system. *Which division is most active in a stress situation? How can you account for this?*

6. Special senses

To illustrate the sensory functions of the nervous system, a study of some special sense organs might be included. The eye and ear can be discussed in some detail, while cutaneous senses might be left to laboratory activities.

A. The Eye

The teacher can illustrate the parts of the human eye and its functions using a take-apart model, or a diagram similar to 25-6; a sheep or cow eye might be obtained from a local butcher shop for dissection. Students may be interested in knowing some causes for nearsightedness, farsightedness, and astigmatism. These can be presented during discussion or as student assigned reports. The teacher may also wish to include a study

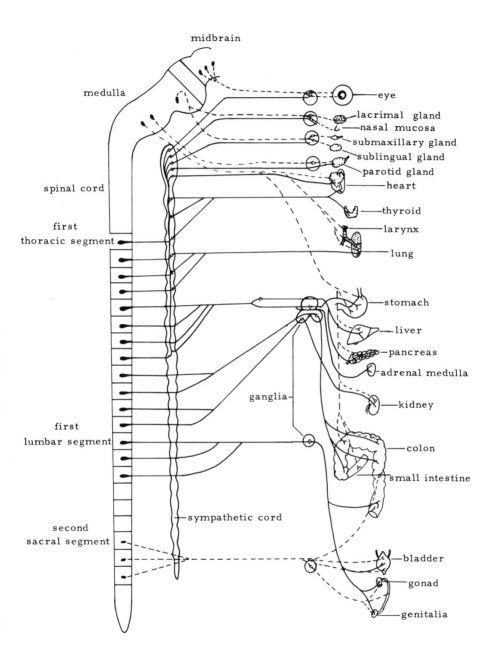

Figure 25-5. Autonomic Nervous System.
*Redrawn from Villee, Walker and Smith,
General Zoology, 3rd ed., Saunders, Philadelphia, 1968*

of the chemical changes which occur in the retina as the result of stimulation by light.

A visit to an eye clinic, optometrist, or school physician might be arranged so that special testing equipment can be demonstrated and observed.

B. Ear

A diagram similar to Figure 25-7, commercial charts, or take-apart models can be used to illustrate the parts of the ear and their function. *How can infection travel from the throat to the middle ear? Why can damage to the ear lead to a disturbance in equilibrium?*

7. Laboratory activities

Using the knowledge gained in this lesson on the action of the nervous system and its parts, the teacher can reinforce and build upon what has been learned with a series of student performed activities on reflexes and sensations in the human. Most of these activities can be performed by the students on themselves, in the space of a single laboratory or lecture period. These activities enable students to study the role of sensory organs in initiating reflex action; the location of various reflex centers in the brain stem and spinal cord; and the physiology of the reflex arc.

Before proceeding with the activities themselves, have the students prepare for them by study in textbooks and references on the physiology of reflexes. Elicit from them the fact that a simple reflex arc involves sensory nerves, the brain or spinal cord as the coordinating center, and motor nerves. Have the students, in their readings, find the answers to these specific questions: *What are proprioceptive impulses? What evidence have you that muscles, tendons, and ligaments are supplied with sensory nerves? What is a tendon reflex? How do they originate normally in our body? What is inhibition? What is muscle tonus?*

The following activities can be performed by students working in pairs, or in groups of three or four. If more than two students are involved, the first can be the subject, the second can perform the test, the third can record the observation(s), and the fourth can provide an analysis of the data.

A. Tendon Reflexes

1) Patellar Reflex

Have a student sit on a table so that the leg from the knee down hangs free (or have student sit with legs crossed). The weight of the lower leg puts the quadriceps femoris muscle under tension. Let another student tap the subject's patellar tendon (just below the knee) with the edge of

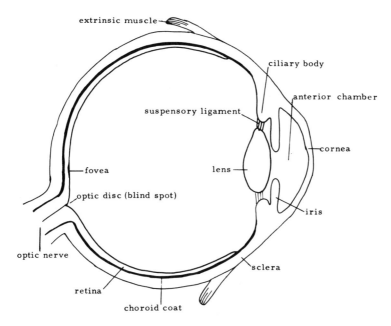

Figure 25-6. Eye.

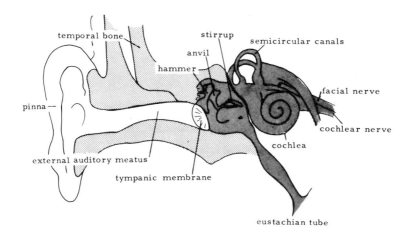

Figure 25-7. Ear.

his hand or a ruler. Try this on three or four subjects. *Is stretching obtained just as easily and is it equally extensive in all subjects? What muscle and tendon are involved in this action? Why must the leg be hanging free in order for this action to occur?* Just previous to the striking of the tendon let the subject clench his fist. *What effect does this have on the knee jerk?* Also give the subject a column of figures to add or some other mental task, and while he is thus engaged, tap the tendon. *Why is there reinforcement of the reflex? How can this be explained in terms of cerebral inhibitions?*

2) ACHILLES TENDON

Have a student kneel on a chair with his feet hanging free over the edge. Let another student gently bend one of the subject's feet forward to increase the pressure on the gastrocnemius muscle and lightly tap the Achilles tendon. *What happens? What muscle and tendon are involved? What role do proprioceptive impulses play in these reflexes?*

B. CUTANEOUS REFLEXES

1) TWO-POINT SENSIBILITY

The subject should be seated with his bare arm resting palm up on the table top. He should have his eyes closed as his partner uses blunt pins to apply two stimuli to the forearm. *What areas are most sensitive to tactile stimulation? How close together must the two points be so that the subject can not distinguish two separate points?* This "two-point threshold" should be recorded for the nape of the neck, forearm, inner lip, a finger, nose, and upper arm.

2) PRESSURE

A student should slowly move one hair across the back of his hand. *Is any sensation apparent? If the hair is moved rapidly is there any sensation?* Other variations of this can be done, including simply pressing a finger rapidly or slowly against the back of the hand.

C. ORGANIC REFLEXES

1) CORNEAL REFLEX

Have a student suddenly bring his hand near his eye. *Why did he blink?* In their readings students may have discovered an interesting medical application of this corneal or wink reflex, since it is practically the last reflex to disappear in anesthesia. *Why is the corneal reflex an important protective response in administering anesthetics?*

2) PHOTO-PUPIL

Have a student close his eyes for two minutes. Then, while facing a bright light, such as a flashlight, he opens them. Another student immediately examines the pupils for changes in size. *What is the object of this reflex? Where is the reflex center for this light reflex located? What muscles are involved? If the light is shined in one eye only, will both irises dilate?*

3) ACCOMODATION

In moderate light, have a student look at an object 20 feet away and then have him shift his gaze to one 10 inches away. Another student examines the pupils for changes in size. *What is the object of this reflex? What role does the lens of the eye play in this reflex? What role do the ciliary muscles play? Why should the light intensity remain constant?*

4) CONVERGENCE

While a subject is looking at a distant object, a student observes the subject's eyeballs and reports on changes in their position. Let the subject look at a near object. *What change in the eyeballs? What muscles bring this about? How is this reflex related to the reflex of accomodation?*

5) CILIOSPINAL

Have a student observe the pupils of a subject's eyes while the subject pinches himself on the back of the neck. *Why do the pupils dilate?*

6) BLIND SPOT

On a large, white index card or sheet of paper, the student should draw a heavy black cross. This should be located in the center of the paper and be approximately ½ inch in height. A minus sign should be drawn 4 inches away, taking care that it is the same size and thickness. Staring at the cross, the student should slowly bring the card toward him. *At what point is the minus sign no longer visible? What causes the blind spot?*

7) AFTER IMAGE

Using a sheet of unlined white paper and one of black, and a small colored card, students can study *after image*. The subject should place the colored card on the black sheet and stare at it until the edges of the card appear to blur. It is important that the eyes do not move at all. When the blurring is noted, the gaze should be moved to the white sheet. *What is seen? What color is the after image? If this is repeated but the subject closes his eyes instead of staring at the white sheet, what color will the image be?*

8) SWALLOWING

Without eating or drinking anything, have a student swallow, then immediately after, have him try to swallow again. Now have the student drink a glass of water, noting the rate at which he can swallow. *How can the differences in rate between swallowing the contents of a glass of water and swallowing without eating or drinking be explained? What parts of the body are involved in swallowing? Where are the reflex centers for swallowing located?*

9) ACUITY

Students can use sterile cotton to plug one of the subject's ears. Bringing a ticking watch near the open ear, the student should move it closer until the subject hears it. Record the distance from the ear to the watch. Repeat for the other ear. *Are both ears equally sensitive?*

10) IDENTIFICATION OF SOUND LOCATION

With eyes closed, the subject should be asked to indicate the location of a ticking watch. *Is he able to locate it best when directly ahead, to left or right, above head, below head, or behind head?*

11) MIDDLE EAR DEAFNESS

Strike a tuning fork and immediately touch the base of the handle to the mastoid bone behind the ear. When vibration is no longer heard, hold the fork immediately outside the ear canal. *Can it still be heard? In middle ear deafness, would air conduction of sound be better, worse, or equal to bone conduction?*

When all of the activities in this lesson have been performed, have the students examine and analyze the results of the activity they conducted. Then have each group write up their particular activity, listing results, conclusions, and inferences, in the form of a scientific paper. One member from each group can report to the class on their particular activity. This serves not only to familiarize the students with all of the activities performed, but also to stimulate class discussion concerning the results obtained and the nature of reflex acts.

PERTINENT FACTS

- A simple reflex arc is composed of sensory and motor nerves, and the brain or spinal cord is the coordinating center.
- Reflex actions differ from such actions as walking or writing in that they are not performed voluntarily; an individual may or may not be aware that a reflex action is occurring.
- A reflex action is purposive and adaptive in that it is conducive to the well-being of the individual.
- During sleep the cortex of the brain is relatively inactive, while all other bodily structures continue to function normally.
- The knee jerk varies in extent in direct proportion to the general mental activity and alertness of the subject.
- As cutaneous receptors are varying distances from each other a certain minimum distance between two stimuli is required in order that they may be identified as two stimuli.
- Under natural conditions, the pupil of the eye is most widely dilated in the dark and most constricted in bright light.
- The object of the photo-pupil reflex is to shield the eye from too great and too sudden illumination and over stimulation of the retina; the center for this reflex is located in the midbrain.

- Adjustment of focus for objects at different distances (accomodation) is necessary because a sharply focused image must fall upon the photoreceptor cells in order to produce clear vision.
- The act of accomodation and the directing of the line of sight (convergence) constitute an associated reflex.
- Areas in the medulla and lower pons of the brain control swallowing.
- Both ears may not be equally sensitive to sound.
- In middle ear deafness, air conduction of sound is less efficient than bone conduction.

POSSIBLE QUIZ

1. Identify the parts of the central and peripheral nervous systems, including the structure and function of neurons.
2. Briefly define the following: afferent fiber, efferent fiber, reflex arc, synapse, somatic, visceral.
3. Discuss the divisions of the autonomic nervous system and their interrelated functions.
4. Explain the functioning of the parts of the eye as a ray of light enters from the surroundings.
5. Trace the path of a sound from the vibrating source to identification by the brain.

READINGS

Baker, B., "Reflexes in the Human Being: Experiments to Propel Study," *Professional Growth for Teachers, Science,* January, 1966.

Eccles, J.C., "The Synapse," *Scientific American,* January, 1965.

Katz, B., "How Cells Communicate," *Scientific American,* September, 1961.

———, "The Nerve Impulse," *Scientific American,* November, 1955.

Marrazzi, A.S., "Messengers of the Nervous System," *Scientific American,* February, 1957.

Snider, R.S., "The Cerebellum," *Scientific American,* August, 1958.

Von Bekesy, G., "The Ear," *Scientific American,* August, 1957.

Wald, G., "Eye and Camera," *Scientific American,* August, 1950.

Walter, W.G., "The Electrical Activity of the Brain," *Scientific American,* June, 1954.

FILMS

"Human Brain." 11 minutes, sound, bw, $2.15. Encyclopedia Britannica Films, Inc., 1150 Wilmette Avenue, Wilmette, Illinois 60091.

"The Multicellular Animal, Part IV: The Nervous System." 23 minutes, sound, color, $8.15. McGraw-Hill Book Co., Text-Film Division, 330 W. 42nd Street, New York, N.Y. 10036.

"The Multicellular Animal, Part IV: The Senses." 30 minutes, sound, color, $8.15. McGraw-Hill Book Co., Text-Film Division.

"Reflexes—Their Principles and Changes in Nervous Diseases." 34 minutes, silent, bw, $4.15. E. Herz, and T.J. Putnam, c/o Columbia University Medical Center, 630 W. 168th Street, New York, N.Y. 10032.

MODELS

1. Using colored clay or plastic, students can construct models of different portions of the brain.
2. Models of the brains of vertebrates, and charts on the nervous system may be obtained from General Biological Supply House, Inc., 8200 South Hoyne Avenue, Chicago, Illinois 60620.
3. Student assembly models and kits on the eye, ear, human brain, and human skull and brain, may be obtained from General Biological Supply House, Inc.
4. Take-apart models of the human eye and ear may be obtained from Ward's Natural Science Establishment, Inc., P.O. Box 1712, Rochester, New York 14603.

Lesson 26

HOMEOSTASIS

Lesson time: 45–90 minutes

AIM

To develop the concept of an internal steady state in man through a study of the interrelations of the major organs and systems of the body.

MATERIALS

Charts and diagrams of the organ systems of man; Film.

PLANNED LESSON

Basing this lesson on the material previously covered in studying the various systems, the teacher can point out the intricate interplay of systems vital to equilibrium within the body. By listing various organs suggested on the chalkboard and quickly reviewing their functions, it will soon become apparent that all organs studied thus far are directly or indirectly involved in maintaining constant conditions in the fluids surrounding the cells. These constant conditions are necessary for the normal growth and functioning of cells. Students should understand that this maintenance of constant conditions is *homeostasis*.

1. The cell

A brief review of the functioning of the cell (Lesson 7) should include the need for glucose, fatty acids, amino acids, water, oxygen, and various ions for a surrounding fluid of a very precise chemical balance. *How are these produced, transported, and controlled?* This will lead directly to a study of the major systems and the role each plays.

2. Circulatory system

Students should be able to identify the transport of materials in the blood as vital to homeostasis—that the diffusion of substances into and out of the blood and fluids surounding the cells insures the constant mixing of all body fluids (Lesson 23). *Why is the constant mixing of fluids important? What substances will be carried by the blood? How does the circulatory system help maintain normal body temperature?* A review of lymph and its functions should also be included at this point.

3. Respiratory system

The concepts of internal and external respiration (Lesson 22) should be reviewed and the transport of gases again related to blood circulation.

4. Digestive system

The chemical breakdown of foods to produce glucose, amino acids, and fatty acids can be mentioned here, as well as the uses of each in the cell. Again, the circulatory system can be cited in discussing the transport of these substances as well as the transport of the wastes of cellular digestion.

5. Endocrine glands

The regulatory mechanisms of hormones can be linked to the functioning and regulation (control) of the other systems and organs. Their role in the conversion of some digestive end products into materials suitable for use within the cell should be stressed. The role of the liver might be mentioned in this capacity as well.

6. Kidneys

The waste filtration and reabsorptive functions (Lesson 21) of these organs should be readily identified by your students as directly involved with the control of the fluids surrounding the cell. Again, the importance of various ions can be illustrated here, as the kidney is the primary control organ of electrolytes.

7. Nervous system

The nervous control of the organ systems, as well as the various special sensory organs important in helping the organism to be aware and survive within his environment, can all be reviewed as a summation of this total interdependence within the body. The control of body temperature by the nervous system should be included in this study. Stress should be placed on the controls exerted by the autonomic nervous system.

This lesson can act as a culmination of the study of the organ systems of man which are involved with the normal maintenance of the organism. It will also serve as a review of body function on which to base a study of the special processes of reproduction (Lesson 27).

PERTINENT FACTS

- The term homeostasis refers to the maintenance of a constant internal state.
- The cell is surrounded by fluid through which all materials diffuse.
- All of the major organ systems of man are involved in maintaining homeostasis.
- The skeletal and muscular systems serve indirectly in the maintenance of homeostasis.

POSSIBLE QUIZ

1. Based on what you have learned regarding the mechanisms of homeostatic control, is one organ system more important than any of the others? Use as much specific information as possible to support your answer.
2. Discuss the homeostatic control of body temperature.
3. Explain the homeostatic control of the autonomic nervous system.
4. Why is homeostasis important?
5. Discuss electrolytic balance within the body.

READINGS

Benzinger, T.H., "The Human Thermostat," *Scientific American*, January, 1961.

Brown, G.S., and D.P. Campbell, "Control Systems," *Scientific American*, September, 1952.

Langley, L.L., *Homeostasis*. New York: Reinhold Publishing Co., 1965.

Mayerson, H.S., "The Lymphatic System," *Scientific American*, June, 1963

Nourse, A.E., *The Body*. New York: Life Science Library, Time, Inc., 1964.

Overmire, T.G., *Homeostatic Regulation*, BSCS Pamphlet Series, No. 9. Boston: D.C. Heath and Co., 1963.

Smith, H., "The Kidney," *Scientific American*, December, 1953.

Zuckerman, S., "Hormones," *Scientific American*, March, 1957.

FILMS

Any of the films listed in Lessons 21, 23, 24, or 25, can be selected and shown for this lesson.

MODELS

Charts and models of the major organ systems of man may be obtained from General Biological Supply House, Inc., 8200 South Hoyne Avenue, Chicago, Illinois 60620; and from Ward's Natural Science Establishment, Inc., P.O. Box 1712, Rochester, New York 14603.

Lesson 27

REPRODUCTION AND DEVELOPMENT

Lesson time: 45–90 minutes

AIM

To develop an understanding of the reproductive system and developmental processes of man.

MATERIALS

Birth models; Embryology models; Films.

PLANNED LESSON

A discussion of the origin of egg and sperm and the development of a new organism should be the concluding lesson in the study of human anatomy and physiology. The functioning organism having been studied, its means of reproducing itself will provide a natural conclusion. Students generally respond well to such a lesson, provided the subject is discussed in a matter-of-fact manner.

1. Origin of reproductive cells

The teacher will need to review the process of meiosis (Lesson 9) so that his students will be thoroughly familiar with the mechanism by which the haploid number of chromosomes is produced.

A. Male

The production of sperm in the testes and their maturation should be explained in terms of the process of meiosis. *What is the importance of the tail on the spermatozoan?* The concept that the motility of the sperm is necessary for fertilization should be included. The spermatozoa and secretion combination which composes *semen* can also be studied.

B. Female

The location and general function of the ovaries as egg producing *gonads* would logically follow here. The production of a *Graafian follicle*, and its subsequent rupture and release of the egg might also be considered. The teacher may also wish to include some information concerning the hormonal control of these structures. The concept that eggs do not have motility should be brought out. The film "The Sex Cells" can be used to present or review this information.

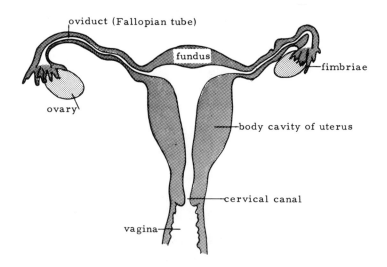

Figure 27-1. Female Reproductive System.

2. Fertilization

The study of fertilization can begin with the movement of sperm after they have entered the female reproductive system. Again, the lack of motility on the part of the egg should be mentioned. The movement of the sperm up into the *oviducts*, as the egg is released from the ovary and moves down the oviduct, can be traced using a diagram similar to Figure 27-1. *What conditions are required in order that the union of gametes occurs?*

Discussion should bring out the need for healthy and fully effective cells, as well as the necessity for the sperm to be in the oviduct when an egg has been released. This will lead naturally into a discussion of the menstrual cycle and ovulation, if the teacher wishes to include it within the lesson. The role of hyaluronidase, the fertilization membrane, and fusion should be included, together with the implantation of the fertilized egg or *zygote*. The film "Fertilization" should prove quite useful in discussing these factors, and should be shown if available.

3. Embryological development

The processes of cleavage, blastulation, and gastrulation might be included at this point. It is important that students understand that, although regular mitotic divisions occur, cells do not separate from each other, nor does cytoplasmic volume increase during early divisions. The teacher might wish to include some information concerning the various historical theories of development at this time.

The film "Theories of Development" can be used to present some early as well as more modern theories on the development of an organism. Using frog development, the film traces a number of interesting theories. Some theories the teacher may wish to present in class would include preformation, epigenesis, and encasement. After this study, the class can move on to a study of the stages of development.

A. Cleavage

A sketch similar to Figure 27-2 can be used to illustrate the division of cells to the *morula* or 32-cell stage. It should be pointed out that the number of cells has increased but that the total volume of the cells is approximately the same or somewhat less than the volume of the original single cell.

B. Blastulation

The migration of cells to form a hollow ball or sphere can be illustrated next. Clay models or commercially prepared models can be used to show the structure of the blastula.

C. Gastrulation

The invagination of the blastula might best be explained by stressing the fact that cells will now begin to increase in size at varying rates. *If only the cells at the top of the blastula increase in length, how will the shape of the hollow ball be changed?* Using this, the teacher can help students to understand the folding-in process. The production of a layered structure can now be seen as well.

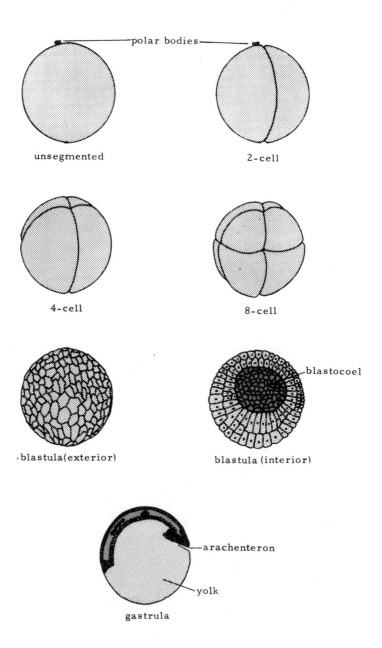

Figure 27-2. Stages of Cleavage and Early Development (Frog).

D. Differentiation

As a study of the movement and specialization of cells would be quite detailed, the teacher can simply indicate that cells are being specialized, and that three layers of cells are involved. The *ectoderm, endoderm,* and *mesoderm,* and the structures which develop from them, can be included.

E. Fetal development

The major changes throughout the stages of development and the formation of the placenta and fetal membranes can be traced. Special stress should be placed on the importance of the *amnion* to man and to other mammals. *What is the function of amniotic fluid? How do food substances and gases pass into the fetus from the blood stream of the mother? How do waste materials leave the fetus? Where do they go?*

The teacher may place the following outline on the chalkboard to indicate the various stages of fetal development:

1. first week—fertilization;
2. second week—implantation;
3. fifth week—limb-buds appear;
4. sixth week—fingers and toes appear, lens of eye forms;
5. eighth week—human appearance;
6. third month—nails form; can determine sex;
7. fifth month—movement of embryo;
8. seventh month—eyes open; weighs approximately 3 lbs.;
9. eighth month—weighs $4\frac{1}{2}$ to 5 lbs.;
10. ninth month—weighs approximately 7–8 lbs., birth.

An excellent film to illustrate fetal development to your students is "Biography of the Unborn."

F. Birth

Diagrams and/or models can be used to illustrate the birth process and to trace the movement of the fetus through the birth canal. *What initiates the birth process? What is the function of oxytocin? Why do infants born prematurely with only six months development usually not survive?*

PERTINENT FACTS

- Gametogenesis involves the process of meiosis.
- Fertilization occurs in the oviduct (Fallopian tube) of the female reproductive system.
- Many sperm must be present in the reproductive tract so that a sufficient quantity of hyaluronidase is produced.
- Hyaluronidase is necessary for sperm penetration of the egg.
- The ectoderm gives rise to the skin, brain and spinal cord, and parts

of the sense organs; the mesoderm to parts of the circulatory, excretory, skeletal, and muscular systems; and the endoderm to the lining of the internal organs, the liver, and the pancreas.
- The placenta and fetal membrane play a major role in the development of the human embryo.
- The birth process is initiated by hormonal and muscular actions.

POSSIBLE QUIZ

1. Explain the development of egg and sperm cells. How does fertilization occur?
2. Trace the development of the fertilized egg through the process of differentiation.
3. Why would damage to the embryo be most likely to occur during the first three months of pregnancy? What effect on the embryo might be expected if the mother contracts measles, is exposed to radiation, or takes harmful drugs during this period?
4. Trace the passage of the fetus through the birth canal during labor.
5. Explain the importance of the amnion, chorion, and allantois.

READINGS

Alen, R.D., "The Moment of Fertilization," *Scientific American*, December, 1950.

Arey, L.B., *Developmental Anatomy: A Textbook and Laboratory Manual of Embryology*. Philadelphia: W.B. Saunders Co., 1965.

Monroy, A., "Fertilization of the Egg," *Scientific American*, December, 1950

Pincus, G., "Fertilization in Mammals," *Scientific American*, March, 1951.

Tyler, A., "Fertilization and Antibodies," *Scientific American*, May, 1955.

Villee, C.A. (ed.), *The Control of Ovulation*. London: Pergamon Press, 1961.

FILMS

"Biography of the Unborn." 17 minutes, sound, bw, $3.90. Encyclopedia Britannica Films, Inc., 1150 Wilmette Avenue, Wilmette, Illinois 60091.

"The Human Body: Reproductive System." 13 minutes, sound, color, $5.40; bw, $3.90. Coronet Films, Coronet Building, Chicago, Illinois 60601.

"Human Reproduction." 23 minutes, sound, bw, $4.40. McGraw-Hill Book Co., Text-Film Division, 330 W. 42nd Street, New York, N.Y. 10036.

"Reproduction, Growth, and Development: Fertilization." 21 minutes, sound, color, $8.15. McGraw-Hill Book Co., Text-Film Division.

"Reproduction, Growth, and Development: The Sex Cells." 17 minutes, sound, color, $8.15. McGraw-Hill Book Co., Text-Film Division.

"Reproduction, Growth, and Development: Theories of Development." 30 minutes, sound, color, $8.15. McGraw-Hill Book Co., Text-Film Division.

MODELS

1. Students can construct clay models of the different stages of embryological development, or these can be obtained from Carolina Biological Supply Co., Burlington, North Carolina 27215.
2. Charts on the human reproductive system as well as birth models may be obtained from General Biological Supply House, Inc., 8200 South Hoyne Avenue, Chicago, Illinois 60620.

Unit V

THE ORIGIN AND EVOLUTION OF LIVING THINGS

Lesson 28

THE ORIGIN OF LIFE

Lesson time: 45 minutes
Laboratory time: 45 minutes

AIM

To introduce a few theories concerning the development of living organisms.

MATERIALS

Jars of preserved specimens of animals of the major phyla; Dissecting trays; Tongs or forceps.

PLANNED LESSON

The lesson can be introduced by the discussion of the possibility of life on other planets. *What are the characteristics of living things? What environmental requirements do living things have?* A discussion of life on other planets can be turned into a discussion of the possible origins of life on the planet Earth. After some general discussion, the teacher might bring in some historical theories, including those of Aristotle, Thales, and Spinoza. This can be followed by some discussion of the currently held theories.

1. Early theories

Student reports would be a good means by which to introduce the various theories held by the early philosophers. The teacher can indicate that all early theoreticians saw the highly specialized and integrated parts and functions of the organism as a result of supernatural creation. This "purposiveness" seen in living things might be illustrated by the students themselves. *Why is it incorrect to assume that an organism has developed a structure because there was a need for it?*

Students must clearly understand that an organism has no control over the code carried by DNA, so that, even if it were conscious of a possible physical adaptation (which would be beneficial), it could do nothing to alter its heredity to insure such a development. Aristotelian "entelechy" might be mentioned at this point. The Empedoclean theory that many highly formed organs developed individually and were then joined to form various types of animals might be mentioned.

The theory that life continuously arose from nonliving matter can be discussed as it affected later thinkers. Some students might enjoy researching the works of these thinkers and reporting on the many "recipes" for producing animals from nonliving substances.

The history of spontaneous generation, including some information about Redi, Leeuwenhoek, and Spallanzani, can be used here to lead into a discussion of Louis Pasteur's experiments which dealt the final blow to the theory. Later research on the nature of organic substances might be used to highlight another of the steps toward a theory of the origin of life.

2. Modern theories

The teacher might have students report on the works of Schroedinger, Ducrocq, Dauvillier, Oparin, Haldane, and many others. The idea that some primitive solution of organic substances, as suggested by Haldane, may have given rise to a protein molecule as the result of lightning excitation can be covered here. The reproductive characteristics of such a material should be explored. The concept that, by accidental changes in the DNA, these structures became more complex, should also be explored. The long process of evolution and the resultant complexity of living things should be discussed. It is also important to consider the interrelations of organism and environment. *How might seriously unfavorable environmental conditions affect the perpetuation of an organism which is a newly developed mutant? How might a new life form alter the environment?*

It should be pointed out to your students that most modern theoreticians now accept Oparin's theory that living matter and nonliving matter are not fundamentally different; that the characteristics of life must have developed in the process of evolution.

The interrelationships between a study of the origin of life and the nature of living things should be stressed. *How does knowledge of the anatomy and physiology of organisms help us to suggest ways in which they may have developed?*

References are available to students who might wish to learn more about the many experiments which have been performed and are being performed, in the study of the origin of life. Newspaper reports of new research might be collected and posted. *Life Begins*, by Moffat and Scheour, might be assigned for outside reading.

3. Evolution

As a natural continuation of the study of the origin of life, the class may move to the process of evolution. Student reports on various early theories of evolution might be used here. Again, Empedocles and Aristotle might be cited. Students should understand that no convincing evidence in support of the theory of evolution was collected until the work of Charles Darwin. Some background concerning Darwin and his trip to the Galapagos Islands can be included, as well as a study of the theory of natural selection, and the origin of species.

The teacher should stress that evolution occurs because of random *mutations* which are advantageous to the organism. Here, the interrelationships between organism and environment should be examined. The teacher might suggest various possible mutations in man. Students should then consider the possible success or failure of the mutant and the reasons for this. The films "From Water to Land" and "The Biochemical Origins of Terrestrial Life" can be used to trace the origin and transition of living things from water to land.

The teacher might briefly mention the major groups of animals, beginning with the one-celled protozoa, pointing out the advances of each group over the previous one. In this manner, the development from single-celled to multicellular, highly complex organisms can be traced. The teacher might introduce some of the theories of classification which are based upon structural similarities and differences.

In the laboratory, students can be given jars which contain a number of different specimens. These jars, or other suitable containers, can be prepared simply by placing preserved specimens of animals of most of the major phyla in them. Included in each jar could be molluscs, annelids, arthropods, echinoderms, roundworms, small fish, reptiles, amphibia, and vertebrates.

Working in groups of four, students should carefully examine the specimens and decide what they have in common and how they differ. Based upon their similarities and/or their differences, the specimens should be arranged in groups. Students should clearly describe the characteristics by which they have identified these groups. Each major group can be divided into as many subgroups as needed in order to classify the specimens. In their laboratory notebooks, students should write out their complete classification scheme, indicating how each specimen has been classified.

As a second activity, each student can be given one specimen, and should then attempt to develop a classification system for that particular specimen. This laboratory activity can be used by the teacher to develop the principles of classification and to introduce the concepts of the classification system of Linnaeus.

The lesson can be concluded with a review of the theories of the origin of life and with the theories of evolution. This will provide a good basis for a study of the theories of the evolutionary development of man.

PERTINENT FACTS

- The theory of spontaneous generation is also known as *abiogenesis*.
- Pasteur did more than any other scientist to place the theory of *biogenesis* on a firm footing.
- Modern theories have been developed around a biochemical approach to the origin of life.
- Animals may be classified according to similarities and differences in structure.

POSSIBLE QUIZ

1. How does the theory of abiogenesis compare with that of biogenesis? Explain fully.
2. Briefly discuss the biochemical origins of life.
3. Using the biochemical theories, suggest a possible pathway of evolutionary development for terrestrial organisms.
4. How would you define the term *life?*
5. What characteristics would you use to classify living things?

READINGS

Adler, I., *How Life Began*. New York: New American Library, 1959.
Asimov, I., *Chemicals of Life*. New York: New American Library, 1962.
Brown, H., "The Age of the Solar System," *Scientific American*, April, 1957.
Gamow, G., *The Creation of the Universe*. New York: Viking Press, 1960.
Moffat, S., and E.A. Schneour, "Continuity and Change," *Life Beyond the Earth*, Vistas in Science, No. 2. Washington, D.C.: National Science Teachers Association, 1965.
Oparin, A., *Life: Its Nature, Origin, and Development*. New York: Academic Press, 1962.

———, *The Origin of Life.* New York: Dover Publications, Inc., 1953.

Reynolds, J.H., "Age of the Elements in the Solar System," *Scientific American,* November, 1960.

Wald, G., "The Origin of Life," *Scientific American,* August, 1954.

FILMS

"Genetics: Biochemical Origin of Terrestrial Life." 30 minutes, sound, bw, $8.15. McGraw-Hill Book Co., Text-Film Division, 330 W. 42nd Street, New York, N.Y. 10036.

"Life, Time, and Change, Part X: Dinosaurs." 24 minutes, sound, color, $8.15. McGraw-Hill Book Co., Text-Film Division.

"Life, Time, and Change, Part X: From Water to Land." 28 minutes, sound, color, $8.15. McGraw-Hill Book Co., Text-Film Division.

"Life, Time, and Change, Part X: Origin of Life." 24 minutes, sound, color, $8.15. McGraw-Hill Book Co., Text-Film Division.

"Prehistoric Times: The World Before Man." 11 minutes, sound, color, $3.40; bw, $2.15. Coronet Films, Coronet Buildings, Chicago, Illinois 60601.

MODELS

A plaque illustrating the evolution of the vertebrate skull may be obtained from Ward's Natural Science Establishment, Inc., P.O. Box 1712, Rochester, New York 14603.

Lesson 29

FROM THE SIMPLE TO THE COMPLEX— EVOLUTION

Lesson time: 45–90 minutes

AIM
To trace the evolutionary development of man.

MATERIALS
Films.

PLANNED LESSON

The teacher can begin the lesson by reviewing some of the historical theories of evolution, with emphasis on those of Darwin. The class should be aware of the role each of the sciences plays in formulating evolutionary theory. The duality of this role, both the development of the theory and the contribution of a great collection of data against which theories can be tested, should be stressed. Students should be aware that evolution by natural selection is a major unifying biological theory.

1. Paleontology

The significance of fossils in the study of man's evolutionary development should be considered. Reference might be made to Erasmus Darwin, Buffon, and Lamarck, as theorists who were all concerned with the fossil record. Special attention should be given to Lamarck's "use-disuse" hypothesis, and to his belief that acquired characteristics could be inherited.

Students should understand that paleontology can not supply information about the interrelationships of animals which lived together, but can only give information about very large periods in time. The fact that paleontology has provided a great deal of information concerning the evolution of man should be mentioned. *What is a fossil? How were they formed?* Reference materials can be used by students to develop an insight into the science of paleontology and the formation and study of the fossil record.

2. Evolution

Charles Darwin's theory of natural selection can be presented here with specific emphasis on the development of man. The total process of the evolutionary development of all living things and their interrelationships might be considered as part of a discussion of the *survival of the fittest*.

The various forms of natural selection can be considered specifically as they relate to man. The concept should be considered that some mutations cause improper functioning or death, and others affect mating and ultimate reproduction either directly by malfunction or indirectly by making the organism in some way unattractive to a potential mate. The variation of environmental conditions as it relates to the success of the organism should also be included, as well as the theory of directional selection (adaptation). Students should understand that organisms may develop alterations of the genetic code in response to environmental conditions, as some genes become less important while others become more important. The films "Darwin and Evolution" and "Natural Selection and Adaptation" might be used here if available.

3. The ancestors of man

The class should be introduced to information concerning the discovery and evaluation of Pithecanthropus, Java Man, Peking Man, Neanderthal Man, Zinjanthropus, Australopithecus, and Homo habilis. If possible, a trip to a nearby museum can be arranged so that students can visit appropriate displays. They can gather information concerning various fossils and their characteristics. *How do these relate to lower forms of life? Is there a continuity in the evolutionary patterns from protozoa through man?* The film "The Evolution of Man" would be of value here.

4. Evolutionary patterns

The teacher might conclude the lesson on evolution by leading a classroom discussion in which the interrelationships of plants and animals throughout the geologic past are considered, as well as the evolutionary changes through which each passed. *Are the kinds of plants and animals living today the same as those of earlier geologic times?* The transition to land forms of early aquatic plant and animal life should be included in the discussion. The film "From Water to Land" is especially well suited to this discussion.

PERTINENT FACTS

- Fossils are important in supplying evidence of evolution.
- Lamarck proposed the theory of use and disuse.
- Charles Darwin developed the theory of natural selection and provided the basis for the modern theory of evolution.
- A mutation is an alteration of the DNA molecule.
- Natural selection provides for evolutionary variety.
- The evolution of all living things—plants, lower animals, and man—is interrelated.
- Evidences of continuing evolution can be seen in insect species which develop resistance to pesticides, and strains of bacteria which become immune to antibiotics.

POSSIBLE QUIZ

1. Explain Darwin's theory of natural selection, and contrast it to Lamarck's theory of inherited characteristics.
2. What are fossils? Why are they important in the study of evolution?
3. Considering man's immediate ancestors, trace his evolutionary development.
4. Discuss the factors influencing the evolution of a species. Why must the interrelationship of environmental factors be considered?
5. What is meant by adaptation? How does man adapt to various environments?

READINGS

Andrews, H., *Studies in Paleobotany*. New York: John Wiley and Sons, Inc., 1961.

Blum, H., *Time's Arrow and Evolution*. New York: Harper Torchbook, Harper and Row, 1962.

Dobzhansky, T., "The Genetic Basis of Evolution," *Scientific American*, January, 1950.

———, "The Present Evolution of Man," *Scientific American*, March, 1960.

Howells, W.W., "The Distribution of Man," *Scientific American*, September, 1960.

Simons, E.L., "The Early Relatives of Man," *Scientific American*, July, 1964.

Stebbins, G.L., *Processes of Organic Evolution*. Englewood Cliffs, New Jersey: Prentice-Hall, Inc., 1966.

Simpson, G.G., *This View of Life: The World of An Evolutionist*. New York: Harcourt, Brace, and World, Inc., 1964.

FILMS

"Life, Time, and Change, Part X: Darwin and Evolution." 30 minutes, sound, color, $8.15. McGraw-Hill Book Co., Text-Film Division, 330 W. 42nd Street, New York, N.Y. 10036.

"Life, Time, and Change, Part X: Evolution of Man." 26 minutes, sound, color, $8.15. McGraw-Hill Book Co., Text-Film Division.

"Life, Time, and Change, Part X: From Water to Land." 28 minutes, sound, color, $8.15. McGraw-Hill Book Co., Text-Film Division.

"Life, Time, and Change, Part X: Natural Selection and Adaptation." 27 minutes, sound, color, $8.15. McGraw-Hill Book Co., Text-Film Division.

"Life, Time, and Change, Part X: Systematics and Plant Evolution." 28 minutes, sound, color, $8.15. McGraw-Hill Book Co., Text-Film Division.

MODELS

1. Charts of geological time tables, and metal casts of prehistoric animals can be obtained from General Biological Supply House, Inc., 8200 South Hoyne Avenue, Chicago, Illinois 60620.
2. A plaque illustrating the evolution of the vertebrate skull can be obtained from Ward's Natural Science Establishment, Inc., P.O. Box 1712, Rochester, New York 14603.

Unit VI

DRUGS AND ADDICTION

Lesson 30

THE MEDICAL USES OF DRUGS

Lesson time: 45 minutes
Laboratory time: 45–90 minutes

AIM
To study the development and uses of drugs.

MATERIALS
Microscopes; Depression slides; Culture of Daphnia; Eye droppers; Antibiotics, antihistamine, amphetamine, aspirin, alcohol, barbiturate, coffee.

PLANNED LESSON
General class discussion will provide a good basis on which to begin the lesson. Students might suggest the names of drugs and the specific uses of each. The teacher may wish to go into some general background information as well.

1. Historical background

If time permits, some discussion of the early practice of medicine should be included. The ancient remedies and practices of Greece, Babylon, and

Egypt can be mentioned. The role of Hippocrates in introducing scientific diagnosis should be stressed. The history of medicine might then be continued through Galen and the alchemists to modern developments. Included should be such researchers as Koch, Lister, Jennings, and Pasteur.

2. World War I to the present

Using the First World War as a focal point, the teacher should indicate the great advances which occurred in the drug field. *Why would the needs of a war provide sudden advances?* Students will need to realize that the United States was not a major manufacturer of drugs, and depended upon the more advanced German firms until that time. *What difficulties had to be overcome?* Some of the research accomplished after this "rebirth" can be discussed in class. These can include Banting and Best's production of insulin, and Minot and Parry's work dealing with the use of liver to treat pernicious anemia. The importance of the work of Domagk in 1937 should be stressed as the beginning of the development of antimicrobial substances.

Students can be asked to prepare reports on various scientists and drugs of importance to man. Included in these reports should be Fleming and his discovery of penicillin, Ehrlich and Waksman and their contributions, and the many other advances which have developed since World War II, such as the Salk vaccine, Sabin, measles, and mumps vaccines.

3. Drugs of importance today

Students should understand that various developments in research have provided many highly specific drugs. Their own experiences can be used to emphasize that a wide variety of medicines are prescribed today. *What advantages might there be in having highly specialized drugs? Why is it advantageous to have several drugs available which perform the same function?*

Wherever possible, the teacher should relate the functions of representative drugs selected for discussion to their effects on the cell or on various organs, as previously presented in the course. Drugs discussed can include various antibiotics, heart stimulants such as digitalis, anticoagulants, antihistamines, pain relievers, and sedatives. Innoculation and vaccination practices can be included as well. The course of such a discussion should be determined by the interests of your class.

4. Laboratory activity

Students can determine the effects of various classes of drugs on the heart and circulatory system, by using an easily observable laboratory animal such as Daphnia (water flea). These can be obtained from commercial suppliers or from tropical fish stores, and should be kept in an isolated aquarium.

The teacher should obtain one sample of a number of different classes of substances, such as a barbiturate, amphetamine, antihistamine, mild narcotic such as found in cough syrup, antibiotic, aspirin, alcohol, and coffee. Drugs which require prescriptions can be obtained from the school or family physician. By preparing water solutions of these in varying strengths, students can investigate the effects of different concentrations of these on the heart.

The Daphnia should be transferred from the culture bottle to a depression slide by means of an eye dropper. There should be only enough water on the slide to keep the Daphnia alive. The Daphnia should then be observed under low power, with special attention being given to the rapidly beating heart. Figure 30-1 can be used to identify the heart of this organism.

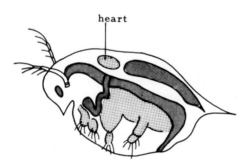

Figure 30-1. Daphnia.

Be sure to time the heartbeat for one minute *before* you add your test solution. *Why?* Place a drop of one of the solutions on the Daphnia. *What is the effect on the heart?* By repeating this procedure with the remaining solutions, the effects of stimulants, depressants, and other materials on the heart can be seen. Each group of students may be assigned a single drug, or if a sufficient number of Daphnia are available, each group may test a number of different solutions. *Why is it necessary to use a fresh specimen for each test? Why are Daphnia well suited to this activity?*

5. The development of a new drug

Tracing the possible stages in the identification and subsequent commercial development of drugs will provide an opportunity to review the scientific method, as well as to illustrate the use of this method in the difficult tasks involved. *Did some unusual occurrence stimulate an investigation? Did the careful observation of an accidental result in the laboratory lead to the discovery? Was an attempt to find a cure for a disease the stimulus? What was the background of the scientist and how did it influence the discovery?*

The stages of the scientific method can be reviewed and discussed as they would apply to a researcher in this field. The painstaking and complex work should be discussed in such a manner that some real understanding can be developed. If possible, a laboratory technician or researcher might be invited to your class to explain the procedures and techniques being followed in some current research.

The lesson can be concluded by emphasizing the great advances which have occured in modern medicine, and the number of drugs that have been developed. Students should be aware that these advances have taken place in a short period of time.

PERTINENT FACTS

- The scientific method in medicine was introduced by Hippocrates.
- World War I marked the beginning of the development of the American drug industry.
- Domagk's work on the sulfonamides was the point from which much research on antibiotics stemmed.
- The work of Paul Ehrlich with his *magic bullet* initiated much of the research for chemotherapeutics.
- In the last decade, science has made tremendous breakthroughs in the development of vaccines for a number of serious diseases, such as polio, measles, and mumps.

POSSIBLE QUIZ

1. Trace the development of the field of medicine from early times to Louis Pasteur.
2. Briefly define the following terms: pharmacology, antibiotic, tranquilizer, depressant, therapeutic, and vaccine.
3. Select one researcher and discuss his contribution to the development of drugs or vaccines.
4. Based on the laboratory activity, what effects do stimulants and depressants have on the heart?
5. What factors can give increased impetus to the development of a new drug or vaccine?

READINGS

Beckman, H., *Drugs: Their Nature, Action, and Use.* Philadelphia: W.B. Saunders Co., 1958.

Cooley, D.G., *The Science Book of Modern Medicines*. New York: Pocket Books, Inc., 1963.

Drill, V.A. (ed.), *Pharmacology in Medicine*. New York: McGraw-Hill Book Co., 1958.

Himwich, H.E. (ed.), *Tranquilizing Drugs*. Washington, D.C.: American Association for the Advancement of Science, 1957.

FILMS

"The Making of a Doctor: A Wider World." 28 minutes, sound, bw, free loan. Merck, Sharp and Dohme Film Library, West Point, Pennsylvania 19486.

"Mission: Measles." 20 minutes, sound, bw, free loan. Merck, Sharp and Dohme Film Library.

"Time of Hope." 20 minutes, sound, color, free loan. Merck, Sharp and Dohme Film Library.

Lesson 31
THE EFFECTS OF DRUG MISUSE

Lesson time: 45–90 minutes

AIM

To study some of the ways in which drugs may prove harmful to man, with emphasis on drug addiction.

MATERIALS

Films.

PLANNED LESSON

The teacher can begin with the more commonly used drugs and the results of their misuse, and continue the lesson through to the addicting drugs and hallucinogens.

1. Non-prescription drugs

Students might list here aspirin, cough medicines, and the many other patent medicines available without a prescription. Accidental overdoses or dosage of the wrong medicine can be considered. *Why must the label on a bottle of medicine always be carefully read? What precautions can be taken to prevent accidents involving drugs? Why is it unwise to take any medication without consulting a physician? Why is it extremely dangerous to use medication originally prescribed for someone else?*

2. Prescribed drugs

Discussion can center around such questions as: *Why should medicines not be saved for future use? Why are prescriptions frequently not re-fillable or may be re-filled only once? Why must the doctor be aware of any medication the patient may be taking for other problems?* Students should be aware of the possible harmful effects drugs may have when taken in combination.

3. Narcotics

The class should first discuss the valuable uses of narcotics in their role as pain killers. The uses of each of the major narcotics should be considered separately.

A. Morphine

The fact that this drug is used as the main pain killer in medical practice should be introduced, as well as the fact that, despite popular belief, this drug does not produce a feeling of happiness and euphoria in healthy people. Its effect in increasing the pain threshold and in reducing pain, as well as its addicting powers, should be discussed as they relate to the need for care on the part of medical practitioners. *What might happen if a patient were given morphine indiscriminately? What is meant by drug dependence?*

B. Codeine

Some discussion of the use of this drug as a cough suppressant should be included. It can be mentioned that, although addicting, large quantities of codeine must be administered before an addiction develops. *Why might cough medicines containing codeine be no longer available unless the purchaser signs for them?*

C. Marijuana

Although actually not a drug, marijuana should be included in the lesson since it is generally involved in the addiction process. The teacher can use this as an example of the many plants which contain potent chemical substances. The danger of smoking marijuana as a step toward drug addiction should be strongly emphasized. *Why would marijuana smoking lead to experimentation with other narcotic substances?* The psychological factors of this transition should be thoroughly explored.

D. Heroin

This more powerful derivative of morphine should be identified as the dangerous drug it is. The ease with which one becomes addicted can be stressed by the fact that, although its pain killing powers require a much smaller dose and it has fewer unpleasant side effects, it is never used in medical practice in place of morphine. The speed with which addiction occurs must be emphasized, so that students realize the dangers involved in its use.

A large variety of literature, free films, and visiting lecturers can usually be obtained through the local narcotics organization in your area. Usually the local police station or hospital can supply you with the names of such groups, as can the Public Health Service and the Bureau of Narcotics of the Treasury Department. The many ramifications involved in heroin addiction should be explored, including the need for larger doses, the high price, and the tragic sociological effects.

4. Stimulants and hallucinogens

Students should be made aware of the many other substances used as stimulants and hallucinogens, including airplane glue.

A. Amphetamines

The functioning of sympathomimetic drugs such as the amphetamines can be deduced by students from the name. From this, they might conclude that drugs which imitate the sympathetic division of the autonomic nervous system (Lesson 25) would have a wide effect on all body functions. Using their knowledge of cell and body function, students should be able to explain that stimulation by amphetamines can be only temporary, as energy production and normal functions have been artificially stimulated, without the addition of all the normal requirements of functioning. The concept of "beating a dying horse" might serve as an apt illustration. The complete exhaustion which follows the depletion of the final reserves of energy should also be discussed.

B. Barbiturates

The depressant effects of these sleep-inducers, as well as their misuse, can be graphically illustrated by pointing to the number of deaths attributed to "overdose of sleeping pills." It should be pointed out that barbiturates are also addicting, and must be used with care. Mention should be made of the results of the combination of barbiturates and alcohol.

C. Hallucinogens

The effects of LSD (d-lysergic acid diethylamide), DMT, mescaline, and psilocybin in the production of hallucinations can best be presented by student reports. This will provide an opportunity for students to read about the peculiar reactions of people who take such drugs. The belief that one can fly, which causes frequent deaths; the euphoria, or more commonly, the horrible visions and fears; and the genetic damage produced, should all be included in these reports. Students should try to locate articles written about specific individual reactions. Numerous films are available for use in the classroom, many of which are listed at the end of this lesson.

As a conclusion to this lesson, the teacher should review the many harmful results of improper drug use, as well as the psychological problems which might lead someone to the use of drugs.

PERTINENT FACTS

- It is important that all drugs be used with care and only under the supervision of a doctor.
- Drugs should never be saved for future use, as every illness should be treated by a physician.
- The misuse of drugs, especially narcotics, can lead to harmful physical and tragic sociological effects.
- The use of hallucinogens, such as LSD, can result in genetic defects as a result of chromosome breakage.
- Amphetamines are known as sympathomimetic drugs because they imitate the sympathetic division of the autonomic nervous system.

POSSIBLE QUIZ

1. How may the accidental misuse of drugs be prevented?
2. Why is it unwise to take any medication without consulting a physician? Why shouldn't drugs be saved for future use?
3. List some major drugs and explain their action.
4. Why must the use of any narcotic be carefully supervised by a physician?
5. What are some of the harmful physical and mental effects of the use of hallucinogens?

READINGS

Ausubel, D.P., *Drug Addiction: Physiological, Psychological, and Sociological Aspects*. New York: Random House, 1964.

Barron, N.F., M.E. Jarvik, and S. Bunwell, "The Hallucinogenic Drugs," *Scientific American*, April, 1964.

Livingston, R.B., *Narcotic Drug Addiction Problems*. Washington, D.C.: U.S. Department of Health, Education, and Welfare, U.S. Government Printing Office.

Nichols, J.R., "How Opiates Change Behavior," *Scientific American*, February, 1965.

United States Treasury Department. *Living Death: The Truth About Drug Addiction*. Washington, D.C.: Bureau of Narcotics, Government Printing Office, 1965.

FILMS

"Drug Addiction." 22 minutes, sound, bw, $4.15. Encyclopedia Britannica Films, Inc., 1150 Wilmette Avenue, Wilmette, Illinois 60091.

"Hooked." 28–31 minutes, sound, bw, free loan. Narcotic Addiction Control Commission, Executive Park South, Albany, New York 12203.

"The Losers." 28–31 minutes, sound, color, free loan. Narcotic Addiction Control Commission.

"Narcotics: Why Not?" 15 minutes, sound, color, free loan. Narcotic Addiction Control Commission.

"The Seekers." 30 minutes, sound, color, free loan. Narcotic Addiction Control Commission.

INDEX

Abiogenesis, 231, 233
Abscission, leaf, 124
Accommodation, 212-213
Acetocarmine stain, preparation, 70
Acetylcholine, 207
Achilles tendon, 160
   reflex, 212
ACTH (adrenocorticotrophic hormones), 200
Actin, 159
Acuity, 213-214
Adaptation, 236
Addiction, drug, 246-248
Addison's disease, 200
Adenine, 62
Adenosine diphosphate (ADP), 172
Adenosine monophosphate (AMP), 172
Adenosine triphosphate (ATP), 56, 172, *173*
ADH (anti-diuretic hormone), 178
ADP (adenosine diphosphate), 172
Adrenaline, 200
Adrenocorticotrophic hormones (ACTH), 200
Adventitious roots, 89
After image, 213
Agglutination, 190, *191*
Alcohol, 242
All-or-none-law, muscle, 159
All-or-none-response, nerve, 207
Alveoli, 181-182
Amino acids, 169, 172
Amnion, 225
Amniotic fluid, 225
AMP (adenosine monophosphate), 172
Amphetamine, 242, 247
Anaphase:
   meiosis, 73
   mitosis, 69
Ancestors of man, 235
Anemia, 189
Antagonist, 160
Anther, 117
Antibiotics, 241
Anticoagulants, 241
Anti-diuretic hormone (ADH), 178
Antihistamine, 242
Antihistamines, 241

Aorta, 194
Apical meristem, 129
Appendix, 168
Aristotle, 230
Arthritis, 157
Aspirin, 242, 245
ATP (adenosine triphosphate), 56, 172, *173*
Atria, 192
Atrio-ventricular node, 193
Australopithecus, 236
Autonomic nervous system, 208, *209*
Auxins, 124-125
   laboratory activities, 130
Axial skeleton, 155
Axons, 206-207

Banting, F.G., 198, 241
Barbiturates, 242, 247
Basophils, 191
Bayliss, Wm., 198
Bean germination, *121*
Benedict's solution, 170
Best, C. H., 198
Bicuspid valve, 192
Biogenesis, 233
Biome, 135
Birth, 225
Birth canal, 225
Birth process, 225
Blade, of leaf, 104
Blastulation, 223
Blind spot, 213
Blood:
   agglutination, 190-191
   clotting, 191-*192*
   plasma, 187-188
   platelets, 191-192
   red blood cells, 188-190
   smear slide, *190*
   typing, 190
   white blood cells, 190-191
Blood cells:
   erythrocytes, 188-190
   leucocytes, 190-191
   platelets, 191
Blood clotting, 191-*192*

INDEX

Blood plasma, 187-188
Blood platelets, 191-192
Blood smear, 189-*190*
Blood typing, 190
Blood vessels, 194-195
Bolus, 167
Bone:
  anatomy, 150
  Haversian canal system, *153*
  long bone, longitudinal section, *152*
Bone marrow, 151
Bouin's fixative, 35
Brain, 203-206, *205*
  cerebellum, 204
  cerebrum, 205
  corpus callosum, 205
  diencephalon, 205
  medulla oblongata, 204
Breastbone, 155
Breathing, 182
  laboratory activity, 182-185, *183*
  physiology, 182-183
Bronchi, 181-182
Bronchioles, 181-182
Bud scar, 98
Buds, 98
Buffon, G., 235
Bundle of His, 193
Bundle scar, 98
Bunion, 157
Bursae, 157
Bursitis, 157

Calyx, 117
Cambium, 99
Carbohydrates, 169
  test for digestion, 169-170
Carnivorous plants (see insectivorous plants)
Carotene, 106
Carpals, 156
Carpels, 117
Carrot roots, 89
Catalyst, 106
Carboxyhemoglobin, 183
Cecum, 168
Cell:
  animal-plant cell differences, 58-59
  blood, 188-192
  cytoplasm, 55-56
  division, 60-75
  endoplasmic reticulum, 57
  DNA, 62, *63*
  estimation of size, 38-39
  estimation of speed, 39-40
  generalized, *47*
  genes, 62

Cell (*cont.*)
  Golgi apparatus, 58
  heredity, 77
  homeostasis, 217-218
  internal structure and function, 55-59
  laboratory activities
    cheek cells, 58
    onion cells, 58
  lysosomes, 57
  membrane, *47*
  mitochondria, 56-57
  nucleus, 61
  phagocytosis, 57
  plate, 69
  reproductive, 221-222
  ribosomes, 57
  RNA, 61
  role in homeostasis, 217-218
  vacuoles, 58
  wall, 69
Cell differences, 58
Cell division, 66-75
  meiosis, 71-75, *72*
  mitosis, 67-71, *68*
    laboratory observation, 70
Cell membrane, 48-54
  diffusion, 48
  osmosis, 48-54
  structure, *47*-48
Cell plate, 69
Cell wall, 69
Central nervous system, 203-206, *204*
  brain, 203-206
  spinal cord, 203
Centrioles, 58
Centromere, 67
Centrosomes, 58
Cerebellum, 204
Cerebral cortex, 205
Cerebral hemispheres, 205
Cerebrospinal fluid, 206
Cerebrum, 205
Chemicals, proper handling:
  dry, 19
  liquid, 19-20
Chemotaxis, 191
Chemotropism, 129
Cholinesterase, 207
Chlorenchyma, 105
Chloroplasts, 58
  laboratory activities, 109
Chlorophyll, 106
Chlorosis, 127
Chromatid, 67
Chromatography, *108*-109
Chromosomes, 61-62, 76-77
  cross, xx, xy, *79*

Chromosomes (cont.)
  distribution of x and y, 78
  genes, 62
Chyme, 170
Ciliospinal reflex, 213
Circulatory system:
  blood, 187-192
  blood vessels, 194-195
  heart, 192-194
  homeostasis, 218
  lymph, 195
Citric Acid Cycle, 172, 173
Classification, principles, 232-233
  laboratory activity, 232-233
Clavicle, 156
Cleavage, 223
  frog egg, 224
Coccyx, 155
Codeine, 246
Codons, 77
Coffee, 242
Cohesion theory, 100
Coleoptiles, 129
Collar bone, 156
Collodion liquid, 48-49
Colloidal solution, 49
  laboratory test, 50
Colon, 168
Communities, 135
Competition, ecological, 136
Conductivity, in nerves, 207
Conductor cells, 128
Controlled experiments, 41-43
Convergence, 213
Corn germination, 120
Corneal reflex, 212
Corolla, 117
Corpus callosum, 205
Cortex:
  kidney, 176
  root, 90
  stem, 100
Cotyledons, 120
Cough medicine, 245
  codeine, 246
Cover slip technique, 33
Cranial nerves, 207
Cranium, 153-154
Crenation, 53
Cretinism, 199
Crick, F.H.C., 62
Cristae, 56
Cutaneous reflexes, 212
Cutin, 100
Cyclosis, 56
Cytokinesis, 69
Cytolysis, 54

Cytoplasm, 55-56
  cytokinesis, 69
Cytoplasmic streaming, 102
Cytosine, 62

Daphnia, 241-242
Dark reaction, in photosynthesis, 106-107
Darwin, C., 232
  theories of evolution, 235
Darwin, E., 236
Dauvillier, 231
Day-neutral plants, 126
Dehydrogenases, 57
Dendrites, 206-207
Deoxyribonucleic acid (DNA), 62-63, 76-77
  replication, 67
Determination of sex, 77-78
Development:
  embryologic, 223-225
  fetal, 225
Diabetes:
  insipidus, 199
  mellitus, 200
Diaphragm, 181-182
Diastole, 193
Dicot, 90
Diencephalon, 205
Differentiation, 225
Diffuse roots, 89
Diffusion, 48-49
Diffusion membranes, 48-49
Digestion:
  uses of end products, 172
Digestive processes, 169-170
  laboratory activities, 170-171
Digestive system, 164-174, 165
  anatomy, 164
  homeostasis, 218
  human, 165
  laboratory activities, 170-171
  mouth, 166
  physiology, 169-172
  tests for
    carbohydrate digestion, 170-171
    fat digestion, 171
    protein digestion, 171
  tooth, 167
  uses of end products of digestion, 172
Digitalis, 241
Dihybrid, 80
Dionaea muscipula, 143-145
Diploid, 71
Disaccharides, 169
DMT, 247
DNA, 62-63, 76-77, 231
  replication, 67

INDEX

Domagk, 241
Dominance, 79, *81*
Drosera rotundifolia, 143
Drug addiction, 246-248
Drug misuse, 245-249
Drugs:
  depressants, 247
  development of, 242
  effects of misuse, 245-249
  hallucinogens, 247-248
  historical background, 240-241
  laboratory activity, 241-242
  medical uses, 240-249
  modern day drugs, 241
  narcotics, 246-247
  non-prescription, 245
  prescription, 246
  stimulants, 247-248
  sympathomimetic, 247
  World War I to present, 241
Duodenum, 168
Dura mater, 203
Ducrocq, 231

Ear, 210, *211*
  laboratory activities, 214
Ecology, 134-141
  communities, 135
  competition, 136
  definition, 135
  ecosystem, 137
  field study, 136-137
  interdependence, 135
  succession, 135
Ecosystem, 137
Ectoderm, 225
Effector cells, 128
Egg osmometer, *52*
Ehrlich, P., 241
Elephantiasis, 195
Embryological development, 223-225
  birth, 225
  blastulation, 223
  cleavage, 223
    of frog egg, *224*
  differentiation, 225
  fetal development, 225
  gastrulation, 223
Empedocles, 231
Endocrine glands, 197-198
Endocrine system, 197-201
  abnormalities, 198
  glands and their functions, 197-198
  hormones and homeostasis, 197-198, 218
  laboratory activities, 198-199
Endoderm, 225

Endodermis, 90
Endoplasmic reticulum, 57
Entelechy, 231
Enzyme, 106
Epicotyl, 120
Epidermis, 90, 92
  leaf, 106
  root, 90
  stem, 100
Epinephrine (see Adrenaline)
Epiphytes, 98
Erythrocytes, 188-190
Esophagus, 168
Estrogen, 199
Evolution, 230-237
  adaptation, 236
  ancestors of man, 236-237
  early theories, 230-231
  Empedoclean theory, 231
  entelechy, 231
  fossils, 236
  genetic changes in DNA, 231
  interrelationships, plant-animal, 237
  laboratory activities, 232-233
  modern theories, 231
  natural selection and origin of species, 232
  paleontology, 236
  of primitive forms of life, 232
  theory of natural selection, 236
  theory of use-disuse, 236
Evolutionary patterns, 237
Excitability and conductivity, 207
Excretion, 177-178
Excretory system, 175-180
  kidney structure
    gross, 176-*177*
    microscopic, 175-176
  laboratory activities
    acetone bodies, 179
    bile, 179
    occult blood, 179
    protein, 179
    specific gravity, 178
    sugar, 178
  nephron, *176*
Exopthalmic goiter, 200
Experiments, controlled, 41-43
Eye, 208-210, *211*
  laboratory activities, 212-213

F.A.A. (formalin, alcohol, acetic acid), 34
Fallopian tubes, 225
False ribs, 155
Fat digestion:
  test for, 171

Fats, 169
Fatty acids, 169, 172
Feedback, 198
Female reproductive system, *222*
Femur, 156
Fertilization:
   in humans, 222
   in plants, 117-119
Fertilization membrane, 223
Fetal development, outline of stages, 225
Fetal membranes, 225
Fetus, 225
Fibrous roots, 89
Fibula, 156
Fixative agents:
   Bouin's, 34-35
   F.A.A., 34-35
Fleming, A., 241
Floating ribs, 155
Flower:
   definition, 117
   parts, 117-*118*
   pistillate, 117
   staminate, 117
Foods, 169
Foramen magnum, 153, 203
Fossils, and evolution, 235-236
Fruits, 119
Funny bone, 151
Furrowing, 69

Galapagos Islands, 232
Gall Bladder, 168
Gametes, 67
Gametogenesis, 225
Gastrulation, 223
Genes, 62, 77
Genetic aberrations, in man, 84-85
   Klinefelter's syndrome, 84
   mutations, 84
   non-disjunction, 84
   Turner's syndrome, 84
Genetic code, 61-65, 236
Genetics:
   aberrations in man, 84
   distribution of X and Y chromosomes, *78*
   dominance, *81*
   laboratory activities, 82-83
   Mendelian pea crosses, *81*
   mutations, 84
   segregation, *81*
Genotype, 79
Geotropism, 129, 132
   laboratory activities, 132

Germination, of seeds, 119
   bean, *121*
   corn, *120*
Gibberellic acid, 125
Glands, endocrine, 198
Glomerulus, 176
Glucagon, 200
Glucose, 172
Glycerol, 169, 172
Golgi apparatus, 58
Gonads, 222
Graafian follicle, 222
Graduated cylinder, proper reading, *20*
Grave's disease, 200
Growth, plant:
   laboratory activities, 129
   plant growth responses, 123-133
   plants and minerals, 131
Growth movements of plants, 123-124
Guard cells, 106, 112
Guanine, 62
Guttation, 94

Haas and Reed's A to Z solution, 131
Haldane, J. B., 231
Hallucinogens, 247-248
   DMT, 247
   LSD, 247
   mescaline, 247
   psilocybin, 247
Hanging drop slides, 33
Haploid, 71
Haversian canal, 151, *153*
Heart:
   anatomy, 192-*193*
   function, 192-193
   innervation, *194*
   systole and diastole, 193
   valves, 192
   ventricles, 192
Heart stimulants, 241
Heart valves, 192
Hemoglobin, 183, 189
Herbaceous stems, 98
Heredity, 76-86
   laboratory activities
      inheritance of a single characteristic, 82-83
      phenotypes, in corn, 82
      phenotypes, in man, 83
Heredity in man, 83-85
Heroin, 246-247
Heterozygous, 79
Hill reaction, in photosynthesis, 106-107
Hippocrates, 241
Homeostasis, 197-198
   function of organ systems in, 217-219

INDEX

Homo habilis, 236
Homozygous, 79
Hormones, 197-200
Human skeleton, *154*
Humerus, 156
Huxley, T.H., 158-159
Hyaluronidase, 223
Hybrid (*see* heterozygous)
Hydroponics, 131-132
Hyperthyroidism, 200
Hypertonic solution, 52, 102
Hypocotyl, 120
Hypothalamus, 205
Hypothyroidism, 200
Hypotonic solution, 53

Ileum, 168
Imbibition, 93
Indeterminate plants, 126
Indoleacetic acid, 130
Inoculation, 241
Insectivorous plants, 142-147
   characteristics, 143-145
   cultivation, 142-143
   laboratory activities, 145-146
Insulin, 200
Intercalated discs, 158
Interdependence, ecological, 135
Internodes, 98
Interphase:
  meiosis, 67
  mitosis, 67
Intestinal juice, 170
Invagination, 223
Isotonic solution, 52

Janus Green B, 57
Java man, 236
Jejunum, 168
Joints, 157
  disorders, 157

Kidney:
  function, 177-178
  glomeruli, 176
  gross structure, 176
  homeostasis, 218
  laboratory activities
    acetone bodies, 179
    bile, 179
    occult blood, 179
    protein, 179
    specific gravity, 178
    sugar, 178
  microscopic structure, 175-176

Kinetochore, 67
Klinefelter's syndrome (see genetic aberrations in man)
Knee cap, 156
Knee jerk, 211
Knop's solution, 131-132
Kreb's cycle, 172-*173*

Laboratory procedures, 18-24
  safety techniques, 19-21
Lacunae, 151
Lamarck, J.B., use-disuse, 236
Lamellae, 151
Large intestine, 168
Law of Dominance, 79
Law of Incomplete Dominance, 80
Law of Independent Assortment, 80
Law of Segregation, 80
Leaf:
  anatomy
    external, 104-105
    internal, *105*-106
  chlorenchyma, 105
  cross section, *105*
  epidermis, 106
  guard cells, 112
  laboratory activities, 114-115
  mesophyll, 105
  midvein, 105
  palisade tissue, 105-106
  parenchyma, 105
  petiole, 105
  phloem, 105
  photosynthesis, 106-107, 113
  respiration, 112-115
  spongy tissue, 106
  stomata, 106, 112-113
  transpiration, 113-114
  vascular bundles, 105
  veins, 105
  wilting, 114
  xylem, 105
Leeuwenhoek, A., 231
Lenticels, 98
Leucocytes, 190-191
Light, effect on plants, 126
Light reaction, in photosynthesis, 106-107
Linnaeus, C., 233
Liver, 168
Lock and Key theory, 106
Locomotion and support, 150-163
Long bone, *152*
Long-day plants, 126
Loop of Henle, 176
LSD (lysergic acid diethylamide), 247
Lung capacity, 184
Lungs, 181-185, *182*

Lymph, 195
Lysergic acid diethylamide (LSD), 247
Lysosomes, 57

Macrosporophyll, 117
Marijuana, 246
Measles vaccine, 241
Measurement:
  microscopic units, 38
Medulla, of kidney, 176-177
Medulla oblongata, 204
Meiosis, 71-75, *72*
Mendel, G., 79
  dominance, 79, 80
  incomplete dominance, 80
  independent assortment, 80
  segregation, 80
Mendelian genetics, 79-82
  crosses, pea, *81*
Mendel's laws, 79-82
Meninges, 203, 206
Meningitis, 203
Meniscus, 20
Menstrual cycle, 223
Meristematic region, 90
Mescaline, 247
Mesoderm, 225
Mesophyll, 105
Metacarpals, 156
Metamorphosis, hormonal effects on, 198-199
Metaphase:
  meiosis, 71
  mitosis, 69
Microns, 39
Microscope:
  care and use, 25-31, *27*
  carrying, 25
  functions, 26
  laboratory activities, 29-30
  parts, 25-28
  resolving power, 26
  units of measurement, 38
Microsporophylls, 117
Middle ear deafness, 214
Midvein, of leaf, 105
Millimeter, 39
Mimosa pudica, 128
Minot, 241
Mitochondria, 56, *57*
Mitosis, 67-71, *68*
Monocytes, 191
Monosaccharides, 169
Morphine, 246
Morula, 223

Motor neuron, *206*
Mouth, *166*
Mumps vaccine, 241
Muscle contraction, 158
Muscle fatigue, 159
  laboratory activities, 161
Muscle system, 158-161
  anatomy, 158
  attachment, 160
  characteristics, 159
  dark and light bands, 158-159
  diseases and disorders, 161
  theories of contraction, 158-159
  tonus, 159
Muscle tonus, 159
Mutations, 84, 232
Myofibrils, 158
Myosin, 159

Narcotics, 246-247
Natural selection, 232
Natural selection, theory, 236
Neanderthal man, 236
Necrosis, in plants, 127
Nephron, *176*
Nerve, impulse conduction, *207*
Nerve tissue, 207
  excitability and conductivity, 207
Nervous system, 202-216
  autonomic, 208-*209*
  brain, 203-206
  central, *204*
  characteristics, 207
  conduction of impulse, *207*
  homeostasis, 218-219
  laboratory activities, 210-215
  motor neuron, *206*
  peripheral, 206-207
  special senses, 208-210
  spinal cord, 203
Neuromyal junction, 207
Neurons, *206*-207
  non-myelinated, 206
  motor, *206*
  myelinated, 206
Neutrophils, 191
Nodes, 98
Nondisjunction, 84
Nuclear membrane, 61
Nucleolus, 61
Nucleoplasm, 61
Nucleus, 61-65

Oparin, A.I., 231
  origin of life theory, 231

INDEX 257

Origin of life:
  evolution, 230-237
  Haldane theory, 231
  Oparin theory, 231
Osmometer:
  egg, *52*
  potato, *51*
Osmosis, 48-54
  demonstrations, 51-52
  in plants, 93, 100
Osmotic solutions, 52-53
Osteocytes, 151
Ovaries:
  flower, 117
  human, 222
Oviducts, 222
Ovulation, 223
Ovule, flower, 117
Oxyhemoglobin, 183
Oxytocin, 225

Pain relievers, 241
Paleontology, 236
Palisade tissue, 105-106
Pancreatin solution, 171
Parasympathetic nervous system, 208
Parathormone, 200
Parenchyma cells, 90, 99, 105
Parry, 241
Pasteur, L., 231
Patella, 156
Patellar reflex, 210
Peking man, 236
Pelvis, 156
Penicillin, 241
Pericardium, 192
Pericycle, 90, 99
Permanent mount slides:
  preparation, 34-36
Periosteum, 151
Peripheral nervous system, 206-207
Peristalsis, 168
Petiole, 105
Phagocytosis, 57, 191
Phalanges, 156
Pharynx, 167
Phenotype, 79
Phenotypic ratio, 82
Phenylthiocarbamide (P.T.C.), 83
Phloem, 90, 105
Phosphate, 62
Photolysis, 107
Photoperiod, 126
Photoperiodism:
  laboratory activities, 130-131
Photophobic plants, 126, 130-131
Photo-pupil reflex, 212

Photosynthesis, 106-107
  dark reaction, 106-107
  formula, 107
  laboratory activities, 107-109
  light (Hill) reaction, 106-107
  photolysis, 107
Phototropism, 124, 129, 132
  laboratory activity, 132
Pia mater, 203
Pistil, 117
Pistillate flowers, 117
Pitcher plant, 143-*144*
  laboratory activity, 145
Pith, 99
Pithecanthropus, 236
Placenta, 225
Plankton, 136
Plant growth, regulation, 124-127
Plant hormones, 124-125
Plant responses, 123-133
  growth movements, 123-124
  laboratory activities, 129-132
  phototropism, 128-129
  tropisms, 128-129
  turgor movement, 127-128
Plasmolysis, 54
Plastids, 56, 58
Platelets, 191
Plumules, 120
Poinsettia, 126
Pollen, 117
Pollen tube formation:
  laboratory activity, 119
Pollination, 118
Polysaccharides, 169
Potato osmometer, *51*
Prehistoric man:
  Australopithecus, 236
  Homo habilis, 236
  Java man, 236
  Neanderthal man, 236
  Peking man, 236
  Pithecanthropus, 236
  Zinjanthropus, 236
Preserved materials, storage, 22
Pressure reflex, 212
Prime mover, 160
Progesterone, 199
Prophase:
  meiosis, 71
  mitosis, 67
Protein digestion:
  test for, 171
Proteins, 169
  test for protein digestion, 171

Protozoa:
  culturing, 22
  estimating speed with microscope, 39
  suppliers, 22
Psilocybin, 247
P.T.C. (phenylthiocarbamide), 83
Pulmonary vessels, 194
Pulvinus, 128
Punnett Square, 78-79
Purkinje fibers, 193

Radius, 156
Ranunculus root, cross section, *91*
Receptacle, flower, 117
Receptor cells, 128
Recessiveness, 79
Rectum, 168
Red blood cells, 188-190
Redi, F., 231
Reflex arc, 203
Reflexes:
  accommodation, 212-213
  Achilles, 212
  acuity, 213-214
  after image, 213
  blind spot, 213
  ciliospinal, 213
  convergence, 213
  corneal, 212
  identification of sound, 214
  middle ear deafness, 214
  patellar, 210-212
  photo-pupil, 212
  pressure, 212
  swallowing, 213
  two-point sensibility, 212
Renal pyramids, 177
Reproduction, *222*
  human, 221-223, *222*
  plant, 116-119
Reproduction and development in man, 221-227
  embryological development, 223-225
  female reproductive system, *222*
  fertilization, 222-223
  sex cells, origin, 221-222
Reproductive cells, 222
Reproductive system, female, *222*
Respiration:
  breathing apparatus, 181
  external, 183-184
  internal, 184
  laboratory activities, 184-185
  in leaves, 112-115
Respiratory system, 181-185
  anatomy, 181
  bell jar apparatus, *183*

Respiratory system (*cont.*)
  homeostasis, 218
  lungs, *182*
  physiology, of breathing, 182-183
Rheo discolor, 114
Ribose nucleic acid (RNA), 63
Ribosomes, 57
RNA (ribose nucleic acid):
  mRNA, 63
  rRNA, 63
  tRNA, 63
Root cap, 89
Root hairs, 92
Root pressure, 93
  observation in laboratory, 93-94
Roots:
  anatomy, 89-92
  cap, 88
  cortex, 90
  economic importance, 95
  endodermis, 90
  epidermis, 90, 92
  functions, 88-89
  guttation, 94
  meristem, 90
  parenchyma, 90
  pericycle, 90
  phloem, 90
  physiology, 93-94
  pressure, *94*
  Ranunculus, cross section, *91*
  region of elongation, 90
  region of maturation, 90
  root hairs, 92
  root pressure, 93-94
  stele, 90
  types, 89
  xylem, 90
Rouleaux arrangement, 189
Rugae, 168

Sabin, A., 241
Sacrum, 155
Salivary glands, 166
Salk, J., 241
Sarracenia purpurea, 143
Scapula, 156
Schroedinger, 231
Scientific method:
  applications, 42-43
  steps, 41-42
Sedatives, 241
Seed:
  germination, 119-120
  parts, 120
Segregation, *81*
Sella turcica, 153

INDEX 259

Semen, 222
Sense organs, 208-210
  ear, 210
  eye, 208-210
Sepals, 117
Sex, determination, 77-78
Sex cells:
  eggs, 222
  sperm, 222
Sex chromosomes, 77-78
Sexual reproduction:
  in humans, 221-226
  in plants, 116-121
Shoot tension, 102
Shoot tension theory, 100
Short-day plants, 126
Shoulder blade, 156
Simple balance, use of, 19
Sino-atrial node, 193
Skeletal system, 150-158, *154*
  bone anatomy, 150-151
  cranium, *154*
  disorders, 157
  dissections and preparations, 157-158
  joints, 157
  organization, 151-156
Skeleton:
  human, *154*
  laboratory activities, 157
Skull, 153
Sleep, 205
Slides, preparation, 32-37
  fixative agents, 34-35
  permanent mounts
    standard procedure, 35-37
  staining, 34
  temporary mounts
    hanging drop, 33
    wet mounts, 33
Small intestine, 168
Sodium chloride solution, 49
Sodium citrate solution, 53
"Sodium pump," 207
Sound location reflex, 214
Spallanzani, L., 231
Special senses, 208-210
  ear, *211*
  eye, 208-210, *211*
  laboratory activities, 210-214
Sperm cell formation, 222
Spermatozoa, 222
Spinal cord, 203
Spinal nerves, 207
Spindle, 67, 69
Spinoza, 230

Spongy tissue, 106
Spontaneous generation, 231, 233
Stamens, 117
Staminate flowers, 117
Starch production, by leaves. 107-108
Starling, E. H., 198
Stele, 90, 99
Stem:
  anatomy
    external, 98-*99*
    internal, 98-100, *101*
  cambium, 99
  cohesion theory, 100
  conduction of fluids, 100-102
  cortex, 100
  cutin, 100
  economic importance, 102
  epidermis, 100
  functions, 97-98
  growth towards sun, *125*
  herbaceous, 98
  kinds, 98
  pericycle, 99
  phloem, 99
  physiology, 100-102
  pith, 99
  shoot tension theory, 102
  stele, 99
  translocation, 100
  vascular bundles, 99
  woody, 98
  xylem, 99
Sternum, 155
Stigma, 117
Stimulants, 247-248
Stomach, 168
Stomata, 106, 112-113
  laboratory activities, 114
Stunted growth, in plants, 127
Style, 117
Succession, 135
Succus entericus, 170
Sundew, 143-*144*, 146
Survival of the fittest, 236
Suspension, 49
Swallowing reflex, 213
Sympathetic nervous system, 208
Sympathomimetic drugs, 247
Symphysis pubis, 156
Synapse, 207
Systole, 193

Tallqvist scale, 190
Tap roots, 89
Teeth, 166

Telophase:
 meiosis, 73
 mitosis, 69
Tendon reflexes:
 laboratory activities, 210-212
Terrarium:
 kinds, 137-138
 preparation, 138-140
Testes, 222
Testosterone, 199
Thalamus, 205
Thales, 230
Theory of origin of species, 232
Thigmotropism, 129
Throat, 167
Thrombocytes, 191-192
Thymine, 62-63
Thyroid stimulating hormone, 199
Thyroxine, 199
Tibia, 156
Tongue:
 laboratory activity, 166
Tonsils, 166
Tooth, *167*
Trachea, 181-182
Translocation, 100
Transpiration:
 laboratory activities, 114
 in leaves, 113-114
Tricuspid valve, 192
Triplet code, 77
Tropisms:
 definition, 128-129
 kinds, 129
Turgor movements, 127-128
Turgor pressure, 54, 127-128
Turk's saddle, 153
Turner's syndrome, 84
Twig scar, 98
Two-point sensibility reflex, 212

Ulna, 156
Ulnar nerve, 151
Uracil, 63
Ureter, 177
Urethra, 177
Urinary bladder, 177
Urine analysis, 178-179
Use-disuse theory, 236
Uvula, 166

Vaccination, 241
Vacuoles, 58
Vagus nerve, 193
Vascular bundles, 105
Vascular cylinder, 90
Vasopressin (ADH), 178
Veins, of leaf, 105
Vena cavae, 174
Ventricles, 192
Venus flytrap, 143-*145*, 146
Villi, 168
Vital capacity, 184

Waksman, S., 241
Water bath, double boiler, *107*
Water percolation, determination, 137
Watson, J.D., 62
Wet mount slides, 33
White blood cells, 190-191
Wilting, 114
Woody stems, 98
Wright's stain, 190

Xanthophyll, 106
Xylem, 90, 99, 105

Zinjanthropus, 236
Zygote, 77, 223
Zygote formation, in plants, 119